AF292965

ROBUST MODAL CONTROL
with a Toolbox for Use with MATLAB®

ROBUST MODAL CONTROL
with a Toolbox for Use with MATLAB®

Jean-François Magni

ONERA-Toulouse
CERT/DCSD
Toulouse, France

Springer Science+Business Media, LLC

MATLAB is a registered trademark of The MathWorks, Inc.

ISBN 978-1-4613-5170-2   ISBN 978-1-4615-0637-9 (eBook)
DOI 10.1007/978-1-4615-0637-9

©2002 Springer Science+Business Media New York
Originally published by Kluwer Academic/Plenum Publishers, New York in 2002
Softcover reprint of the hardcover 1st edition

http://www.wkap.nl/

10 9 8 7 6 5 4 3 2 1

A C.I.P. record for this book is available from the Library of Congress

Contents

Acknowledgments xi

Introduction 1

Outline 5

1. MODAL CONTROL: A TUTORIAL 9

  1.1 Generalities and guide-lines for choosing eigenstructure to be assigned 11

    1.1.1 List of symbols and abbreviations 12

    1.1.2 System notations 13

    1.1.3 Eigenstructure notations 14

    1.1.4 Assignable eigenstructure 15

    1.1.5 Definition of the concept of mode 17

    1.1.6 Eigenvector assignment and decoupling 18

    1.1.7 Projection of open-loop eigenvectors 22

    1.1.8 Minimum energy assignment 24

  1.2 Eigenstructure assignment I: Traditional approaches 27

    1.2.1 State feedback or proportional output feedback 28

    1.2.2 Pole assignment by output feedback 32

    1.2.3 Insensitive state feedback design 34

    1.2.4 Non-interactive control design 37

    1.2.5 Dynamic feedforward 42

    1.2.6 Dynamic extension for output feedback design 45

    1.2.7 Observer definition 47

    1.2.8 Observer-based output feedback 54

| | 1.2.9 | Observer-based state feedback | 58 |

1.3	Eigenstructure assignment II: Multi-model approaches	63
	1.3.1	Introduction	64
	1.3.2	Dynamic feedback with given structure	67
	1.3.3	Definition of a frequency domain template	68
	1.3.4	Multi-model eigenstructure assignment	69
	1.3.5	Multi-model "phase control"	70
	1.3.6	Controller order reduction	72
	1.3.7	Illustrative example: multi-model assignment	73
	1.3.8	Illustrative example: multi-model "phase control"	74

1.4	Eigenstructure assignment III: Feedback gain tuning	79
	1.4.1	Introduction	80
	1.4.2	Constraints for pole shifting and gain structuring	82
	1.4.3	Examples of files to be written by the user	88
	1.4.4	Constraints for more advanced objectives	90

1.5	Modal analysis of a control law	95
	1.5.1	Modal simulation and residuals	96
	1.5.2	Modal controllability	99
	1.5.3	Zeros and "almost zeros"	102
	1.5.4	Miscellaneous tools	103

2.	SOME CONTROL DESIGN PROBLEMS	107
2.1	Recommendations and proposed design cycle	109
	2.1.1	General recommendations	109
	2.1.2	Multi-model sub-toolbox	111
	2.1.3	Tuning sub-toolbox	112
	2.1.4	Proposed design cycle for direct feedback design	113
	2.1.5	Improvement of an existing feedback law	115

2.2	Single-model pole placement	117
	2.2.1	From open-loop to closed-loop	118
	2.2.2	Closed-loop eigenvalue insensitivity	122
	2.2.3	Tuning of a dynamic controller	124

| 2.3 | Decoupling | 129 |
| | 2.3.1 | Non-interactive control by proportional feedback | 130 |

2.3.2 Non-interactive control by dynamic feedforward   134

2.3.3 Observer-based non-interactive control   135

2.3.4 Illustration of Exact Loop Transfer Recovery   137

2.4 Multi-model eigenstructure assignment   139

2.4.1 Multi-model design with a proportional gain   140

2.4.2 Multi-model design with a dynamic gain   142

2.4.3 Multi-model design using the tuning procedure   145

2.5 Flexible systems control   149

2.5.1 Low and high frequency feedback designs   150

2.5.2 Single step dynamic feedback design   152

2.5.3 Generalized phase control   153

2.5.4 Iterative technique   155

2.5.5 Technique based on controllability analysis   157

2.6 Structured gain computation   159

2.6.1 Structured dynamic feedback   160

2.6.2 Controller order reduction   162

3. TOOLBOX REFERENCE   167

3.1 List of available functions   169

3.2 Help messages   173

3.2.1 Function: ADD_CSTR   174

3.2.2 Function: ADD_DYN   175

3.2.3 Function: ADD_OBS   176

3.2.4 Function: AZER   178

3.2.5 Function: CHOL_EV   179

3.2.6 Function: CLEAN_EV   180

3.2.7 Function: COMP_DV   181

3.2.8 Function: CONTR_EV   183

3.2.9 Function: CRIT_CTR   183

3.2.10 Function: CRIT_K   185

3.2.11 Function: CSTR_DP   186

3.2.12 Function: CSTR_EIG   188

3.2.13 Function: CSTR_EV   190

3.2.14 Function: CSTR_INI   192

3.2.15 Function: CSTR_K 192

3.2.16 Function: CSTR_QUD 193

3.2.17 Function: DEFIN_VW 195

3.2.18 Function: DEMODATA 198

3.2.19 Function: DFB_INS 198

3.2.20 Function: DFB2OBS 199

3.2.21 Function: DFB_PROJ 201

3.2.22 Function: DIST_QUD 203

3.2.23 Function: DP_CSTR 204

3.2.24 Function: DYN2STA 205

3.2.25 Function: EIG_CSTR 206

3.2.26 Function: EIG_FB 207

3.2.27 Function: FB_DYN 208

3.2.28 Function: FB_PROP 209

3.2.29 Function: FB_TUN 211

3.2.30 Function: FB_VIEW 212

3.2.31 Function: FF_ASSGN 213

3.2.32 Function: FF_STAT 214

3.2.33 Function: KTF_CRIT 215

3.2.34 Function: KTF_CSTR 216

3.2.35 Function: LSIM_MOD 217

3.2.36 Function: OB_GENE 219

3.2.37 Function: OB_INS 221

3.2.38 Function: OBS2DFB 223

3.2.39 Function: PLOT_CON 224

3.2.40 Function: PLOT_RES 226

3.2.41 Function: PLOT_ZER 228

3.2.42 Function: RCAMDATA 229

3.2.43 Function: RCAMPOLE 231

3.2.44 Function: RCAMSTEP 232

3.2.45 Function: SFB_INS 233

3.2.46 Function: SFB_PROJ 236

3.2.47 Function: SOB_PROJ 237

3.2.48 Function: SORT_EV 239

3.2.49	Function: STA2DYN	239
3.2.50	Function: STR_CSTR	240
3.2.51	Function: STR_VIEW	242
3.2.52	Function: VICIN_EV	243

4. APPENDIX 1: PROOFS OF THE RESULTS STATED IN THE FIRST CHAPTER — 245

4.1	Results relative to eigenvector assignment	248
	4.1.1 Proof of Lemmas 1.1.1 and 1.1.2	248
	4.1.2 Proof of Lemma 1.2.4	249
	4.1.3 Proof of Lemma 1.3.1	250
	4.1.4 Linear Quadratic Programming	251
4.2	First order perturbations	255
	4.2.1 Proof of Lemmas 1.2.2 and 1.4.1	255
	4.2.2 Proof of Lemma 1.4.2	256
	4.2.3 Proof of Lemma 1.3.2	259
	4.2.4 Linear Quadratic Programming	260
4.3	Minimum energy assignment	262
	4.3.1 The Hamiltonian solution to the Linear Quadratic Problem	262
	4.3.2 Proof of Lemma 1.1.3	263
4.4	Pole assignment by output feedback	265
	4.4.1 Technical preliminary results	265
	4.4.2 Proof of Lemma 1.2.1	266
4.5	Non-interactive control design	268
4.6	Dynamic feedforward design	274
	4.6.1 Rendering eigenvalues non controllable	274
	4.6.2 Eigenvector pseudo-assignment	276
	4.6.3 Proof of Lemma 1.2.3	277
4.7	Observers	278
	4.7.1 Elementary observers: Proof of Lemma 1.2.5	278
	4.7.2 Observer with Kalman filter structure: Proof of Lemma 1.2.7	280
	4.7.3 Observer transfer function matrix: Proof of Lemma 1.2.6	281

4.7.4	Separation Principle: Proof of Lemma 1.2.8	282
4.7.5	Equivalent dynamic controller	284

5. APPENDIX 2: ADDITIONAL TOPICS 289

5.1 Models used for demonstrations 289

    5.1.1 A simplified flexible aircraft 289

    5.1.2 A bank of linearized models of an aircraft (the RCAM) 290

5.2 Matrices CSTR and CRIT 295

5.3 Installation and system requirements 297

Conclusion 299

References 303

About the Author 309

Index 311

Acknowledgments

The experience behind this work was acquired during several years spent at ONERA Toulouse. The proposed tools have been tested and improved through being used as the design tools for some applied research contracts supported by DRET, STPA, SPAé, GARTEUR, BriteEuRam. I would like to thank my former doctoral students or colleagues who intensively used the proposed software tools as they were being developed: Aka Manouan, Tania Livet, Yann Le Gorrec, Carsten Döll and Caroline Chiappa. Finally, as most of the work devoted to writing this book and to the software development was spent during evenings, weekends and holidays, I would like to thank my relatives for their understanding and patience.

ROBUST MODAL CONTROL
with a Toolbox for Use with MATLAB®

Introduction

Robust control theory has received much attention in the past decades. The most popular approaches are based on frequency domain analysis and design. Early work in this field can be found in [Rosenbrok, 1974], or [Horowitz, 1991]. The most representative modern frequency domain techniques are Loop Transfer Recovery [Doyle and Stein, 1979, Maciejowski, 1989], $\mathcal{H}_\infty$-synthesis [Doyle et al., 1989, McFarlane and Glover, 1990], μ-synthesis [Zhou and Doyle, 1999] and their LMI-based (Linear Matrix Inequality) extensions. Unfortunately, these techniques are not very efficient for problems involving robustness with respect to *real* parameter variations. Alternative techniques which are more appropriate for dealing with *real* uncertainties have also been considered, for example: quadratic stability [Garcia et al., 1997, Khargonekar et al., 1990], adaptation of μ-synthesis [Young, 1996] value set and parameter space techniques [Barmish, 1994, Ackermann, 1993] and so on. But for various reasons (such as: conservatism – special requirements relative to the dependency on uncertainties – limitations on the number of uncertainties – very large controller orders) these techniques do not really solve the general problem. As we shall see in this book, multi-model control design based on modal approaches, despite some more or less heuristic features, is very efficient, especially in the case of real uncertainties.

Some readers might be amazed to see the words "robust" and "modal" associated in a title. Modal control, *i.e.*, pole assignment, eigenstructure assignment, design techniques of the Geometric Approach ([Wonham, 1979]), is often assumed to lead to poor robustness. This belief is mostly due to the development, during the seventies - eighties, of unusable pole assignment techniques which aimed at assigning the poles of multi-input systems without any additional design objectives. Indeed, pole assignment relative to a multi-input system is not unique, in fact an infinity of solutions exist. It is not very realistic to expect good performance and robustness after having ignored part (often the main part) of the available design freedom. This problem was not encountered when single-input systems were considered on account of the uniqueness of the solution.

The Geometric Approach was a first attempt to take into account the full design freedom beyond pole assignment. But, again, in most cases, a part of this freedom was still ignored. It is in fact a very closely related approach, namely the "eigenstructure assignment" approach, that allows the designer to take advantage of all the design freedom. Several techniques for good selection of assigned eigenvectors from allowable subspaces have been proposed. For example, it was proposed to minimize pole sensitivity with respect to small parameter variations. An alternative possibility consisted of shaping the dynamic responses by assigning eigenvectors satisfying some decoupling properties. Although no mathematical evidence was available, controller designed using such techniques were seen *a posteriori* to naturally exhibit good robustness properties (see for example the flight control systems of some military and civil aircrafts in current operation).

This book and the accompanying toolbox treat a selection of standard eigenstructure assignment techniques. Several new or less common features of modal control are also presented. For example:

- Concerning eigenvector selection, it is suggested to use preferably orthogonal projections of open-loop eigenvectors rather than decoupling ideas. A new concept of minimum energy assignment is also proposed.

- Concerning robustness, this book goes farther than the aforementioned natural degree of robustness as it enables the designer to synthesize multi-model control laws. Multi-model control design is an original way to treat robustness with respect to *real* uncertain parameters without conservatism.

- Some classical ideas are also revisited:

  - For example, an original treatment of observers is proposed: unknown input observer design becomes straightforward and observer-based feedback can be synthesized with very low orders.

  - Conversion from standard dynamic controllers to observer-based controllers is also treated.

  - Pole/zero cancellation is rephrased in a multivariable setting *i.e.*, in addition to poles, eigenvectors are also taken into account.

- Phase control is also rephrased in a multivariable and multi-model setting. In fact, the concept of "phase" is replaced by closed-loop eigenvector properties.

- The proposed multi-model tools can also be used for re-designing a given control law (for example obtained by $\mathcal{H}_\infty$ or μ-synthesis). This technique consists of re-assigning the closed-loop dominant eigenvectors assigned by the given controller. Several advantages of considering such a modal version of a given controller such as: order reduction, controller design with a given structure and multi-model robustness improvement will also be discussed. This approach will also be shown to be a powerful way for combining the complementary advantages of frequency domain and modal approaches.

The accompanying toolbox contains three groups of functions. The first group corresponds to standalone functions (see §1.2 and 1.5). For example, some functions permit the user to design a controller just by defining settling time and damping ratio requirements. Analysis tools, especially for dominant pole analysis, belong also to this group. Besides these more or less standard tools, there are two subsets: the *multi-model sub-toolbox* (§1.3) and the *tuning sub-toolbox* (§1.4). The multi-model sub-toolbox permits the designer to choose the structure of the considered dynamic gain with possible frequency domain constraints. Such a gain can be designed in order to assign a selected eigenstructure of one or more models and/or to perform multi-model "phase control". The tuning sub-toolbox is a set of tools that must be combined in a short MATLAB-file that can be used, for example, for shifting poles of one or more models towards given areas of the complex plane.

This book/toolbox is suited for use by students, researchers and engineers in the field of control engineering, who have some acquaintance with linear system theory. For users with little experience in modal control design, a set of basic recommendations are listed at the beginning of Chapter 2. This list of recommendations is followed by the description of a general control design procedure including modal analysis, worst case analysis, single and multi-model synthesis.

Outline

Chapter 1. This chapter gives an overview of basic theory for eigenstructure assignment. Proofs are not given.

- Section 1.1 explains the classical ideas for selecting eigenvectors to be assigned. Alternative ideas are also proposed as projections and "minimum energy pole placement".

- Section 1.2 presents standard modal control design techniques, including non-interactive control and insensitivity. Pole/zero cancellation and observers are presented in an original way.

- Section 1.3 presents the main contribution of this book, *i.e.*, multi-model dynamic eigenstructure assignment. Phase-control is also rephrased in a multivariable and multi-model setting. Using the same formulation, it becomes possible to consider frequency domain constraints and to use structured dynamic controllers. The theory presented in this section can also be used for combining frequency domain and modal techniques and for controller order reduction.

- Section 1.4 presents an iterative technique for improving a given feedback law. The proposed technique consists of solving sequentially a Linear Quadratic Programming problem that, for example, aims at "pushing" some closed-loop poles towards one or more given areas of the complex plane. The same idea can also be used for eigenvector tuning.

- Section 1.5 is devoted to analysis tools. In particular, dominant mode analysis is addressed.

The design techniques and the use of the accompanying toolbox are illustrated by short academic examples.

Chapter 2. The use of the toolbox is illustrated considering more complex and more detailed examples than in the previous chapter. The user can investigate the toolbox starting from this chapter while keeping an eye on Chapter 1 for theory and on Chapter 3 for more details on software tools. The first section of this chapter gives a list of recommendations and a design cycle is proposed. Then the following problems are addressed:

- Section 2.2 presents pole placement. In particular a systematic strategy, based on open-loop eigenvector projection, is proposed. Insensitivity is also considered.

- Section 2.3 is devoted to non-interactive control design. The standard approach is presented first. Alternative approaches based on pole/zero cancellation and observer-based controllers are also treated. Exact Loop Transfer Recovery is presented as an unknown input observer design problem.

- Section 2.4 presents multi-model control design.

- Section 2.5 illustrates the various ways modal control can be used for flexible structure control. In particular, the efficiency of the multivariable generalized phase control technique is demonstrated.

- Section 2.6 illustrates the derivation of the equivalent modal version of a given controller. This "modal translation" can be performed with order reduction or structuring of the feedback gain.

Chapter 3. This chapter provides details concerning the use of the MATLAB functions. In particular, a conventional description of the correct command line usage for optional arguments is specified on page 173. Note that illustrative examples that can be run by "copy and paste" are available in the `help2` message of most functions. There are three subsets of functions:

- Standalone functions. These functions correspond to analysis and single-model design.

- Functions of the multi-model sub-toolbox. These functions *must be combined*. In order to reduce the complexity, it was preferred to split the design function into more elementary functions, each one depending on a single model. These functions are `fb_dyn` and all other functions with names of the form `..._cstr` or `..._crit`.

- Functions of the tuning sub-toolbox that *must be combined* into an iterative m-file to be written by the user. These functions are `fb_tun`

and all other functions with names of the form **cstr_..** or **crit_....**
Before using these functions, read first §1.4.

Chapter 4. This chapter is an appendix giving the proofs of the results stated in Chapter 1. It can be read almost independently of other chapters.

Chapter 1

MODAL CONTROL: A TUTORIAL

Contents.

- Notations and generalities .... page 11
- Choice of eigenstructure .... page 18
- Assignment of eigenstructure .... page 27
- Multi-model case .... page 63
- Tuning of a control law .... page 79
- Modal analysis .... page 95

1.1 Generalities and guide-lines for choosing eigenstructure to be assigned

Contents. The following points will be addressed.

- Which properties has eigenstructure to satisfy to be assignable and which constraints has a feedback gain to satisfy for assigning eigenstructure? .... §1.1.4

- Concept of mode .... §1.1.5

Then, it is shown how design objectives depend on closed-loop eigenstructure selection (eigenstructure assignment will be considered later in §1.2, §1.3 and §1.4).

- Eigenvector selection for decoupling:

  - setting to zero some entries of v_i or w_i. .... §1.1.6
  - vectors in the kernel of some matrix. .... §1.1.6

- Selection from open-loop properties:

  - projection of right open-loop eigenvectors .... §1.1.7
  - minimum energy assignment .... §1.1.8

References. Some ideas presented in this section were used by many authors. For decoupling ideas, the initial contribution is [Moore, 1976], see also the references in [Liu and Patton, 1998]. One of the most convincing application can be found in [Farineau, 1989]. The use of projections as presented here is extracted from [Le Gorrec et al., 1998b, Magni et al., 1998].

1.1.1   List of symbols and abbreviations

$*$	:	Convolution product
$\mathbb{R}, \mathbb{C}$	:	fields of real and complex numbers
$\Re, \Im$	:	real and imaginary parts
j	:	$j = \sqrt{-1}$ or an integer number
s	:	Laplace variable
$\boldsymbol{x}, \boldsymbol{u}, \boldsymbol{y}, \boldsymbol{\xi}, \ldots$	:	small italic bold letters denote signals
$u, v, \xi, \eta \ldots$	:	small italic letters denote vectors, except
$i, \ldots, n, p, q$	:	integer numbers
λ, π, β	:	scalar complex numbers
ε	:	scalar real number
$A, U, X, \Pi \ldots$	:	capital letters denotes matrices, except
Δ	:	stands for "variation of"
A, a, b	:	typewriter type style used for software
X_k or x_k	:	kth column (or row, see context) of the matrix X
$\boldsymbol{x}_k, x_k$	:	kth entry of the signal $\boldsymbol{x}$ or of the vector x
$X^T, X^*, \overline{X}$	:	transpose, conjugate transpose and conjugate of X
$\mathrm{Im}X$	:	column span of the matrix X
$\mathrm{Ker}X$	:	null span of the matrix X
$\mathrm{Diag}(A_1, A_2, \ldots)$	:	diagonal or bloc-diagonal matrix
$\mathrm{Row}(A)$	:	$= [A_1\ A_2 \ldots]$ where A_i denotes the ith row of A
I_n, I	:	$n \times n$ identity matrix or of size given by the context
$0_{m,n}, 0$	:	$m \times n$ null matrix or of size given by the context
$[x_1, \ldots, x_q]$	:	matrix the columns of which are $x_1, \ldots, x_q$
MIMO	:	Multi-Input Multi-Output
SISO	:	Single-Input Single-Output

Systems and feedback

$\boldsymbol{x}, \boldsymbol{u}, \boldsymbol{y}$	:	state, input, output signals
A, B, C, D	:	matrices defining the system $\dot{\boldsymbol{x}} = A\boldsymbol{x} + B\boldsymbol{u}$ and $\boldsymbol{y} = C\boldsymbol{x} + D\boldsymbol{u}$
$S(s)$	:	transfer matrix of the system $S(s) = C(sI - A)^{-1}B +$
n, m, p	:	integers s.t. $A \in \mathbb{R}^{n \times n}$, $B \in \mathbb{R}^{n \times m}$, $C \in \mathbb{R}^{p \times n}$
A_c, B_c, C_c, D_c	:	matrices defining a dynamic feedback
$K, G(s)$	:	proportional and dynamic feedback gains
n_c	:	order of the dynamic feedback
A_f, B_f, C_f, D_f	:	matrices defining a dynamic feedforward
$\boldsymbol{z}, \boldsymbol{z}_R$	:	regulated output and reference input
$\boldsymbol{\xi}, \boldsymbol{\xi}_i$	:	vector of the n modes and the ith mode

$\lambda_i, v_i, u_i, w_i, t_i$	:	eigenvalue, right eigenvector, left eigenvector, input and output directions
V, W, Λ	:	matrices of right eigenvectors / input directions $AV + BW = V\Lambda$ and $K(CV + DW) = W$
$V(\lambda_i), W(\lambda_i)$	:	(columns of) are the assignable v_i, w_i

Observers

$z, \hat{z}$	:	vector of estimated signals and its estimate
u_i, t_i, π_i	:	elementary observer of $z = u_i x$: $\dot{\hat{z}} = \pi\hat{z} + (u_i B + t_i D)u - t_i y$
U, T, Π	:	matrices defining an observer of $z = Ux$: $\dot{\hat{z}} = \Pi\hat{z} + (UB + TD)u - Ty$ and $UA + TC = \Pi U$
K_y, K_z	:	observer-based feedback gains $u = K_y y + K_z \hat{z}$
Q_u, Q_y, Q_z	:	Matrices combining inputs, outputs and estimated signals
η	:	$\eta = Q_u u + Q_y y + Q_z z$
$\hat{\eta}$	:	$\hat{\eta} = Q_u u + Q_y y + Q_z \hat{z}$
$U(\pi_i), T(\pi_i)$	:	(rows of) define the elementary observers

1.1.2 System notations

Let us consider a multivariable linear system:

$$\dot{x} = Ax + Bu \qquad (1.1)$$
$$y = Cx + Du$$

where x is the state vector, u is the input vector, y is the measurement vector. It will be assumed that the system has,

$$
\begin{array}{ll}
n \text{ states} & x \in \mathbb{R}^n \\
m \text{ inputs} & u \in \mathbb{R}^m \\
p \text{ outputs} & y \in \mathbb{R}^p
\end{array}
$$

It is also assumed that the system is controlled by a static output feedback and that an input reference is shaped by a feedforward gain H. So the feedback is:

$$u = Ky + Hz_R \qquad (1.2)$$

where z_R stands for the reference inputs. If $D = 0$:

$$\dot{x} = (A + BKC)\, x + BHz_R$$

or if $D \neq 0$ (it is assumed that $(I - KD)$ is non-singular), from (1.1) and (1.2) we have $u = K(Cx + Du) + Hz_R$ or $u = (I - KD)^{-1}KCx + (I -$

$KD)^{-1}Hz_R$, therefore

$$\dot{x} = (A + B(I - KD)^{-1}KC)\,x + B(I - KD)^{-1}Hz_R$$

In a similar way it can be checked that the following form is equivalent

$$\dot{x} = (A + BK(I - DK)^{-1}C)\,x + B(I - DK)^{-1}Hz_R$$

The matrix $\widehat{K} = (I - KD)^{-1}K$ is a feedback gain equivalent to K, but considering the system without its direct transmission D. This is an interesting result as it permits us to ignore the matrix D when it induces cumbersome notation: If D was ignored for deriving some result relative to a feedback gain $\widetilde{K}$, this result is valid for the system with direct transmission provided that $\widetilde{K}$ is replaced by the equivalent gain K. It is straightforward to check that (assuming that $(I + D\widehat{K})$ is non-singular)

$$K = (I + D\widehat{K})^{-1}\widehat{K} \Leftrightarrow \widehat{K} = (I - KD)^{-1}K$$

1.1.3　　Eigenstructure notations

The following notations will be used. The *eigenvalues* of the matrix $A + BK(I - DK)^{-1}C$ will be denoted

$$\lambda_1, \ldots, \lambda_n$$

the *right eigenvectors*

$$v_1, \ldots, v_n$$

and the *input directions*[1]

$$w_1, \ldots, w_n$$

where (definition):

$$w_i = (I - KD)^{-1}KC\,v_i \Leftrightarrow w_i = K(Cv_i + Dw_i) \qquad (1.3)$$

The *left eigenvectors* of the matrix $A + BK(I - DK)^{-1}C$ will be denoted

$$u_1, \ldots, u_n$$

and the *output directions*

$$t_1, \ldots, t_n$$

where (definition)

$$t_i = u_i BK(I - DK)^{-1} \Leftrightarrow t_i = (u_i B + t_i D)K \qquad (1.4)$$

Hypothesis 1. It is assumed throughout (book and toolbox) that assigned eigenvalues are all distinct.

Matrix Notations. For $q \le n$ vectors:

$$\Lambda = \mathrm{Diag}\{\lambda_1, \ldots, \lambda_q\}$$

$$V = [\ v_1 \quad \ldots \quad v_q\ ]\ ;\ W = [\ w_1 \quad \ldots \quad w_q\ ];$$

$$U = \begin{bmatrix} u_1 \\ \vdots \\ u_q \end{bmatrix}\ ;\quad T = \begin{bmatrix} t_1 \\ \vdots \\ t_q \end{bmatrix}.$$

The vectors u_i and v_i are normalized in such a way that

$$UV = I \quad \text{and} \quad U(A + BK(I - DK)^{-1}C)V = \Lambda \tag{1.5}$$

Hypothesis 2. If λ_i is non-real, it is assumed that for some i', $\lambda_{i'} = \overline{\lambda}_i$ is also assigned. In addition, in the matrices V and W, $v_{i'} = \overline{v}_i$, $w_{i'} = \overline{w}_i$ and in the matrices U and T, $u_{i'} = \overline{u}_i$, $t_{i'} = \overline{t}_i$. (It will be shown pages 30 and 49 that V, W, U and T satisfying this hypothesis can be transformed to real matrices whenever it is necessary.)

1.1.4   Assignable eigenstructure

From the definitions of input and output directions ((1.3) and (1.4)) we have:

LEMMA 1.1.1 *A triple (v_i, w_i, λ_i) can be assigned by proportional output feedback if and only if it satisfies*

$$[\ A - \lambda_i I \quad B\ ] \begin{bmatrix} v_i \\ w_i \end{bmatrix} = 0 \tag{1.6}$$

This triple is assigned in closed-loop by any gain K satisfying the linear equation

$$K(Cv_i + Dw_i) = w_i \tag{1.7}$$

Proof. See Appendix 4, page 248. By duality,

LEMMA 1.1.2 *A triple (u_i, t_i, λ_i) can be assigned by proportional output feedback if and only if it satisfies*

$$[\ u_i \quad t_i\ ] \begin{bmatrix} A - \lambda_i I \\ C \end{bmatrix} = 0 \tag{1.8}$$

This triple is assigned in closed-loop by any gain K satisfying the linear equation

$$(u_i B + t_i D)K = t_i \qquad (1.9)$$

When several vectors are considered simultaneously, using the matrix notations of page 15, Equation (1.7) can be written:

$$W = K(CV + DW) \qquad (1.10)$$

and Equation (1.9) can be written:

$$T = (UB + TC)K \qquad (1.11)$$

Depending on the rank of $CV + DW$ or $UB + TC$ the feedback gain K that perform simultaneous assignment can be computed from (1.10) (or (1.11)) with minimum norm, or least squares approximation, see §1.2.1 for details. Equations (1.6) and (1.8) have also an equivalent matrix form that is respectively

$$AV + BW = V\Lambda$$

$$UA + TC = \Lambda U$$

Degrees of freedom beyond eigenstructure assignment. It is interesting to define the vector spaces in which closed-loop eigenvectors can be chosen. From (1.6) and (1.8) these vector spaces are respectively the column span[2] of $V(\lambda_i) \in \mathbb{R}^{n \times m}$ and the row span of $U(\lambda_i) \in \mathbb{R}^{p \times n}$ defined by[3]:

$$\left[\begin{array}{cc} A - \lambda_i I & B \end{array} \right] \left[\begin{array}{c} V(\lambda_i) \\ W(\lambda_i) \end{array} \right] = 0 \qquad (1.12)$$

The vector $\eta_i \in \mathbb{C}^m$ can be viewed as a vector of degrees of freedom for defining v_i and w_i satisfying Equation (1.6):

$$v_i = V(\lambda_i)\eta_i \;\; ; \;\; w_i = W(\lambda_i)\eta_i$$

The dual case is:

$$\left[\begin{array}{cc} U(\lambda_i) & T(\lambda_i) \end{array} \right] \left[\begin{array}{c} A - \lambda_i I \\ C \end{array} \right] = 0 \qquad (1.13)$$

So for some p-dimensional row vector η_i :

$$u_i = \eta_i U(\lambda_i) \;\; ; \;\; t_i = \eta_i T(\lambda_i)$$

1.1.5 Definition of the concept of mode

For simplifying notations, it is assumed that the considered system has no direct transmission ($D = 0$). Let us consider a regulated output z, a reference input z_R, a disturbance d and an initial condition x_0. System (1.1) becomes:

$$\begin{aligned} \dot{x} &= Ax + Bu + E'd \\ y &= Cx + F'd \\ z &= Ex + Fu \end{aligned} \qquad (1.14)$$

1.1.5.1 Definition

Here, the matrix U of page 15 is assumed to have n rows (then from (1.5), $U = V^{-1}$). Let us consider the change of state space basis

$$\xi = Ux \qquad (1.15)$$

where

$$\xi = \begin{bmatrix} \xi_1 \\ \vdots \\ \xi_n \end{bmatrix}$$

The components of this vectors will be called the *modes*.

1.1.5.2 From excitations to modes

The input u of (1.14) is as in (1.2). The effect of the initial condition is modeled by a Dirac delta function $x_0 \, \delta$. So,

$$\dot{x} = (A + BKC)\,x + BH\,z_R + (E' + BKF')\,d + x_0\,\delta$$

i.e.,

$$\dot{x} = (A + BKC)\,x + f$$

where f summarizes all the excitations acting on the system ($f = BHz_R + (E' + BF'K)d + x_0\delta$). After changing the base ($\xi = Ux$)

$$\dot{\xi} = \Lambda\xi + Uf$$

It follows that

$$\xi(t) = e^{\Lambda t} * Uf(t)$$

where "$*$" is the convolution product and $e^{\Lambda t}$ is the diagonal matrix $\mathrm{diag}(e^{\lambda_1 t}, \ldots, e^{\lambda_n t})$. Therefore

$$\xi_i(t) = e^{\lambda_i t} * u_i f(t) = \int_0^t e^{\lambda_i(t-\tau)} u_i f(\tau) d\tau. \qquad (1.16)$$

1.1.5.3   From modes to states

Coming back to the original basis, (1.15) leads to ($V = U^{-1}$) $x = V\xi$ i.e.,

$$x = \sum_{i=1}^{n} \xi_i v_i \tag{1.17}$$

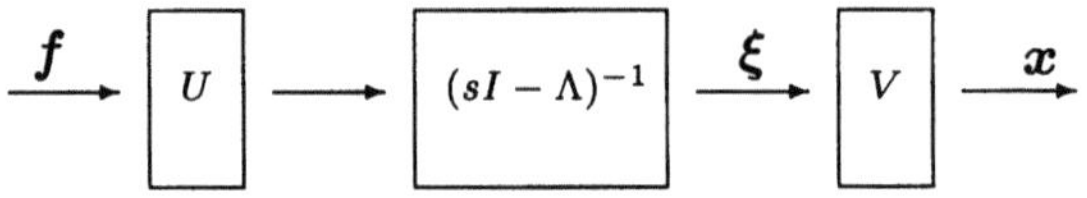

$$\xrightarrow{\;f\;} \boxed{U} \longrightarrow \boxed{(sI - \Lambda)^{-1}} \xrightarrow{\;\xi\;} \boxed{V} \xrightarrow{\;x\;}$$

Figure 1.1.   Block diagram in the modal basis.

1.1.5.4   Summary

Equations (1.17) and (1.16)

$$\xi_i = e^{\lambda_i t} * u_i f \quad \text{and} \quad x = \sum_{i=1}^{n} \xi_i v_i$$

summarize the transfers from the excitation to the modes and from the modes to the state vector. These transfers can be illustrated by the block diagram of Figure 1.1.

1.1.6   Eigenvector assignment and decoupling

This section details modal analysis of the time responses given in §1.1.5. First is treated the response of the regulated output z versus a reference input z_R. Responses to initial conditions and disturbances follow easily. For simplifying notations, it is again assumed that $D = 0$.

1.1.6.1   From a reference input to regulated outputs

Here, $f = BHz_R$ (see §1.1.5). The regulated output $z = Ex + Fu$ is considered. From (1.17), the term Ex can be written $EV\xi$. The feedback signal $u = KCx$ can be written

$$u = KCV\xi$$

and from (1.10) (case $D = 0$)

$$u = W\xi \quad (= \sum_{i=1,\dots,n} w_i \xi_i) \tag{1.18}$$

therefore, Fu can be written $FW\xi$. The transmission of modes is summarized by

$$z = [\ E\ \ F\ ] \begin{bmatrix} V \\ W \end{bmatrix} \xi(t) + \quad \text{direct transmission}$$

more precisely

$$\xi_i(t) = e^{\lambda_i t} * u_i BH z_R \ \text{ and }$$
$$z = \sum_{i=1}^{n} [\ E\ \ F\ ] \begin{bmatrix} v_i \\ w_i \end{bmatrix} \xi_i(t) + \quad \text{direct transmission} \qquad (1.19)$$

The direct transmission term $BHFz_R$ is neglected as it is not related to the modes, so, the transfer from z_R to the modes (ξ) and from the modes to z can be illustrated as follows:

$$\xrightarrow{z_R} \boxed{UBH} \longrightarrow \boxed{(sI - \Lambda)^{-1}} \xrightarrow{\xi} \boxed{[E\ F] \begin{bmatrix} V \\ W \end{bmatrix}} \xrightarrow{z}$$

In order to illustrate the conditions that must satisfy the eigenvectors for decoupling, let us denote E_k, F_k the kth row of E, F; z_k and z_{Rk} the kth entries of z and z_R; H_k the kth column of H:

$$u_i BH_k = 0 \Rightarrow z_{Rk} \text{ has no effect on the mode } \xi_i(t).$$
$$E_k v_i + F_k w_i = 0 \Rightarrow \text{ the mode } \xi_i(t) \text{ has no effect on } z_k.$$

1.1.6.2   From initial conditions to regulated outputs

In the previous case, the system excitation was $f = BH z_R$, here it is replaced by $f = x_0 \delta$. Therefore, the result is similar. After substitution in Equation (1.19)

$$\xi_i(t) = e^{\lambda_i t} * u_i x_0 \delta \ \text{ and }\ z = \sum_{i=1}^{n} [\ E\ \ F\ ] \begin{bmatrix} v_i \\ w_i \end{bmatrix} \xi_i(t) \qquad (1.20)$$

so, the transfer from initial condition to the regulated outputs can be illustrated as follows

$$\xrightarrow{x_0 \delta} \boxed{U} \longrightarrow \boxed{(sI - \Lambda)^{-1}} \xrightarrow{\xi} \boxed{[E\ F] \begin{bmatrix} V \\ W \end{bmatrix}} \xrightarrow{z}$$

The corresponding constraints on eigenstructure for decoupling with respect to initial conditions are summarized as follows

$$u_i x_0 = 0 \Rightarrow \text{ the initial condition has no effect on the mode } \xi_i(t).$$
$$E_k v_i + F_k w_i = 0 \Rightarrow \text{ the mode } \xi_i(t) \text{ has no effect on } z_k.$$

1.1.6.3   From disturbances to regulated outputs

It is assumed that $F = 0$ or $F' = 0$ in order to avoid direct transmissions. Here, $\boldsymbol{f} = (E' + BKF')\boldsymbol{d}$, so, after substitution in Equation (1.19)

$$\boldsymbol{\xi}_i(t) = e^{\lambda_i t} * u_i(E' + BKF')\boldsymbol{d} \quad \text{and} \quad \boldsymbol{z} = \sum_{i=1}^{n} \begin{bmatrix} E & F \end{bmatrix} \begin{bmatrix} v_i \\ w_i \end{bmatrix} \boldsymbol{\xi}_i(t)$$

But, $u_i(E' + BKF') = u_i E' + u_i BKF' = u_i E' + t_i F'$, so,

$$\boldsymbol{\xi}_i(t) = e^{\lambda_i t} * [u_i \; t_i] \begin{bmatrix} E' \\ F' \end{bmatrix} \boldsymbol{d} \quad \text{and} \quad \boldsymbol{z} = \sum_{i=1}^{n} \begin{bmatrix} E & F \end{bmatrix} \begin{bmatrix} v_i \\ w_i \end{bmatrix} \boldsymbol{\xi}_i(t) \quad (1.21)$$

The transfer from the disturbance $\boldsymbol{d}$ to the regulated outputs is illustrated below

$$\boldsymbol{d} \longrightarrow \boxed{[U\ T]\begin{bmatrix} E' \\ F' \end{bmatrix}} \longrightarrow \boxed{(sI - \Lambda)^{-1}} \xrightarrow{\boldsymbol{\xi}} \boxed{[E\ F]\begin{bmatrix} V \\ W \end{bmatrix}} \xrightarrow{\boldsymbol{z}}$$

The corresponding constraints on eigenstructure for decoupling with respect to disturbances are summarized as follows

$$u_i E'_k + t_i F'_k = 0 \Rightarrow \boldsymbol{d}_k \text{ has no effect on the mode } \boldsymbol{\xi}_i(t).$$
$$E_k v_i + F_k w_i = 0 \Rightarrow \text{ the mode } \boldsymbol{\xi}_i(t) \text{ has no effect on } \boldsymbol{z}_k.$$

1.1.6.4   Example

In order to illustrate the decoupling properties that can be expected, let us consider the graph of Figure 1.2.

$$\begin{bmatrix} \boldsymbol{z}_1 \\ \boldsymbol{z}_2 \\ \boldsymbol{z}_3 \end{bmatrix} = \begin{bmatrix} E_1 \\ E_2 \\ E_3 \end{bmatrix} x$$

The decoupling properties illustrated in this graph are:

- the first mode has no effect on $\boldsymbol{z}_3$.

- the third mode has no effect on $\boldsymbol{z}_1$.

- the reference input $\boldsymbol{z}_{R1}$ has no effect on the third mode.

Therefore, we have the following constraints:

$$E_3 v_1 = 0$$
$$E_1 v_3 = 0$$
$$u_3 B H_1 = 0$$

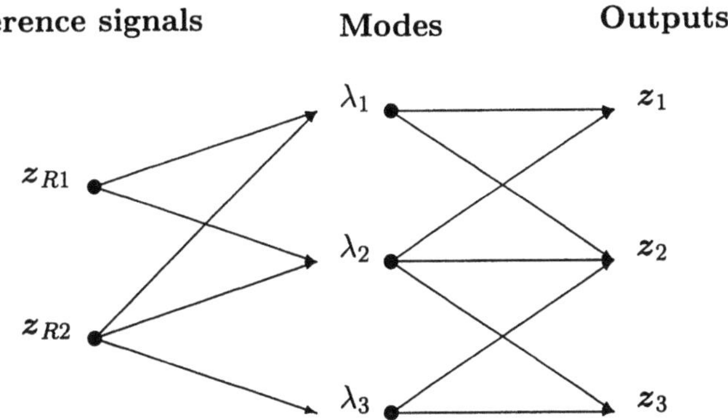

Figure 1.2.   Example of input/mode and mode/output decoupling.

Both first equations should be considered as constraints relative to the feedback gain (K), while the third equation should be considered as a constraint relative to the feedforward gain (H). A more realistic decoupling problem is considered in §1.2.4.

1.1.6.5   Summary of the encountered constraints.

Assigned *right eigenvectors* satisfy (1.6) and decoupling conditions of the form $E_0 v_i + F_0 w_i = 0$. Therefore, v_i, w_i are often computed by solving :

$$\begin{bmatrix} A - \lambda_i I & B \\ E_0 & F_0 \end{bmatrix} \begin{bmatrix} v_i \\ w_i \end{bmatrix} = 0 \tag{1.22}$$

(software option "null space" (key = 'n')). In some cases, each row of the matrix $[E_0 \; F_0]$ has only one nonzero entry, in this case decoupling consists of setting to zero up to $m - 1$ entries of v_i and/or w_i (software option "set to zero" (key = 'z')).

The constraints may also be *over-specified*. For example, assume that the considered system has two inputs. Therefore only one entry of the eigenvectors can be exactly set to zero. The desired eigenvector might be expected to have the following form

$$v_{id} = \begin{bmatrix} * \\ 0 \\ 0 \\ * \\ 0 \end{bmatrix}$$

in which "$*$" denotes non-constrained entries, this specification can also be written as in (1.22)

$$E_0 v_i = 0 \quad \text{where} \quad E_0 = \begin{bmatrix} 0 & 1 & 0 & 0 & 0 \\ 0 & 0 & 1 & 0 & 0 \\ 0 & 0 & 0 & 0 & 1 \end{bmatrix} , \ F_0 = 0$$

so, a least squares solution must be considered. (In the software, the least squares solution is automatically considered for over-specified problems.)

Assigned *left eigenvectors* satisfy (1.8) and decoupling conditions of the form $u_i E_0' + t_i F_0' = 0$. Therefore, u_i, t_i can be computed by solving:

$$\begin{bmatrix} u_i & t_i \end{bmatrix} \begin{bmatrix} A - \lambda_i I & E_0' \\ C & F_0' \end{bmatrix} = 0 \tag{1.23}$$

┌─────────────────────┐
│   Software   │
└─────────────────────┘

Related MATLAB-**functions** (additional details given in §1.2): `defin_vw`, `fb_prop` and `ff_assgn`. These functions consider right eigenvector assignment.

For these functions the user has to choose design options by selecting key:

- `key = 'z'` to set to zero some entries of v_i and w_i.

- `key = 'n'` to have $E v_i + F w_i = 0$.

Examples: See pages 34, 46, 134, 136, and the on-line `help2` message of `fb_prop`.

1.1.7   Projection of open-loop eigenvectors

In order to keep the closed-loop behavior similar to the open-loop one, the most natural strategy consists of shifting to the left badly located *eigenvalues* of a minimum amount, and to assign the orthogonal projection of the corresponding open-loop *eigenvector*. In the same spirit, for well located open-loop eigenvalues, open-loop eigenvalues and right eigenvectors are re-assigned as in open-loop. This strategy will be discussed on page 30 and illustrated on page 118.

Let us denote (λ_{i0}, v_{i0}) a pair of open-loop eigenvalue / eigenvector. An assignable triple (λ_i, v_i, w_i) satisfies (1.6). Let us define what is the assignable triple corresponding to the projection of v_0. Using the matrices $V(\lambda_i)$, $W(\lambda_i)$ defined in (1.12):

$$[\ A - \lambda_i I \quad B\ ]\begin{bmatrix} V(\lambda_i) \\ W(\lambda_i) \end{bmatrix} = 0$$

the projection is characterized by $\eta_i \in \mathbb{C}^m$ where

$$\eta_i = (V^T(\lambda_i)V(\lambda_i))^{-1}V^T(\lambda_i)v_{0i} \qquad (1.24)$$

and

$$v_i = V(\lambda_i)\eta_i \qquad (1.25)$$

$$w_i = W(\lambda_i)\eta_i \qquad (1.26)$$

The benefits of assignment by projection are not limited to preserve open-loop behavior. This kind of assignment is also the key point in order to insure "continuity" in "tuning" a existing feedback design as mentioned later (see page 30). In this case, the reference eigenvectors for assignment by projection are the one assigned by the existing feedback. An other alternative application of projection is proposed in §1.3.6 (page 72). Note that it is also possible to consider in the same way the projection of any desired vector v_{0i}.

```
┌─────────────────────────┐
│        Software         │
└─────────────────────────┘
```

Related MATLAB**-functions** (additional details given in §1.2): `defin_vw`, `fb_prop` and `ff_assgn`. These functions consider right eigenvector assignment. The design options for these functions are:

- `key` = `'p'` projection of an open-loop eigenvector

- `key` = `'v'` projection of a given desired vector.

The functions `dfb_proj`, `sfb_proj` and `sob_proj` use internally the option `'p'`.

The advantage of option `'p'` with respect to option `'v'` is that it is not necessary to compute explicitly the desired vector as it can be identified as the open-loop (or closed-loop) eigenvector corresponding to an eigenvalue. Therefore it suffices to enter the concerned open-loop (or closed-loop) eigenvalue. The option `'p'` permits the designer to assign by projection a closed-loop eigenvector assigned by an alternative gain, but that is possible only if this gain has the same number of inputs and outputs as the one that is being designed, or if it is a state feedback (see the on-line `help1` message of `defin_vw` for more details).

Examples: See pages 119, 140, 143 and the on-line `help2` message of `fb_prop`.

1.1.8   Minimum energy assignment

The minimum energy assignment problem consists of minimizing a quadratic criterion of the form

$$J = \int_0^\infty \boldsymbol{u}^T(\tau)\boldsymbol{u}(\tau)d\tau$$

LEMMA 1.1.3 *The minimum energy eigenstructure assignment corresponds to*

- *preserving the open-loop pairs of eigenvalue/eigenvector for stable open-loop eigenvalues*

- *assigning the symmetric stable images (symmetric with respect to the imaginary axis) of the unstable open-loop eigenvalues. In that case the assigned right eigenvectors are given by:*

$$v_i = (A - \lambda_i I)^{-1} B B^T u_{0i}^*$$

where λ_i is the symmetric stable image of the open-loop eigenvalue λ_{0i} and u_{0i} is the left eigenvector of A corresponding to λ_{0i} (i.e., $u_{0i} A = \lambda_{0i} u_{0i}$).

Proof. See Appendix 4, page 262.

It is not interesting to use this lemma in order to assign *all* eigenvectors because in this case it is better to use directly the LQ approach. We suggest to assign *only some relevant* eigenvalues / eigenvectors in that way. Let λ_{0i} and u_{0i} be a pair of open-loop eigenvalue and left eigenvector, the assigned triple is (λ_i, v_i, w_i).

Rigorous application of the lemma. Applying Lemma 1.1.3 exactly as it is stated means that if λ_{0i} is *unstable*, the assigned eigenvalue is $\lambda_i = -\Re(\lambda_{0i}) + j\Im(\lambda_{0i})$.

Non-rigorous application of the lemma. This result was also applied in a less rigorous way, by shifting to the left some *stable* open-loop eigenvalues. It was often checked that the feedback signals were even smaller than in the projection approach of §1.1.7. So, *it is recommended to use "minimum energy assignment" even for arbitrary pole shifting.*

In both cases, the assigned eigenstructure is given by $v_i = (A - \lambda_i I)^{-1} B B^T u_{0i}^*$ that is:

$$v_i = V(\lambda_i)\eta_i \tag{1.27}$$

$$w_i = W(\lambda_i)\eta_i \tag{1.28}$$

where

$$\eta_i = B^T u_{0i}^* \tag{1.29}$$

and where $V(\lambda_i)$ and $W(\lambda_i)$ have the special form (see the definition (1.12)):

$$V(\lambda_i) = (\lambda_i I - A)^{-1} B \ ; \ W(\lambda_i) = I$$

Software

Related MATLAB-**functions** (additional details given in §1.2): **defin_vw**, **fb_prop** and **ff_assgn**. These functions consider right eigenvector assignment. The design option for these functions is:

- **key** = **'m'** mimimum energy assignment.

Examples: See the on-line **help2** message of **fb_prop**.

1.2 Eigenstructure assignment I: Traditional approaches

Contents. Traditional assignment techniques are considered here. Some classical ideas (*e.g.*, feedforward and observer design) are revisited.

- State or proportional output feedback ....§1.2.1

- Pole assignment by output feedback ....§1.2.2

- Insensitive control design ....§1.2.3

- Non-interactive control design ....§1.2.4

- Dynamic feedforward design ....§1.2.5

- Elementary dynamic output feedback ....§1.2.6

- Observer definition ....§1.2.7

- Observer based dynamic output feedback ....§1.2.8 and 1.2.9

References. Eigenstructure assignment by proportional feedback was first introduced in [Kimura, 1975]. In this reference, some of the right eigenvectors are selected in order to assign more than p eigenvalues (see Lemma 1.2.1). This technique was simplified in the state feedback case in [Moore, 1976]. It is in this reference that modal shaping of regulated outputs (see §1.1.6) appeared first. Most of the work that followed these seminal references is reported in [Liu and Patton, 1998]. An overview of pole assignment is proposed in [Magni and Champetier, 1991]. The translation of the sensitivity of eigenvalues in terms of eigenstructure comes from [Moore and Klein, 1976] and was revisited in [Kaustky and Nichols, 1990, Kaustky et al., 1985]. This section presents non-interactive control design. The relevant references are cited at the beginning of §2.3. Observers are also considered. This concept was introduced by Luenberger in [Luenberger, 1966]. The point of view adopted here is closer to the one in [Wonham, 1970]. The reader is referred to [Magni, 1996, Magni and Mouyon, 1991, Magni and Mouyon, 1994] for more details. The dynamic feedforward treatment of §1.2.5 comes from [Magni, 1987].

1.2.1   State feedback or proportional output feedback

In this section is treated the problem of assigning q triples (λ_i, v_i, w_i) satisfying (1.6). This equation can be written

$$Av_i + Bw_i = \lambda_i v_i$$

The corresponding set of q equations can be written

$$A[v_1 \ldots v_q] + B[w_1 \ldots w_q] = [v_1 \ldots v_q] \operatorname{Diag}\{\lambda_1, \ldots, \lambda_q\} \qquad (1.30)$$

or in matrix form

$$AV + BW = V\Lambda$$

The corresponding set of constraints relative to any gain that assigns these q triples is a set of linear equations similar to (1.7) that can be written globally as

$$K(C[v_1 \ldots v_q] + D[w_1 \ldots w_q]) = [w_1 \ldots w_q] \qquad (1.31)$$

or in matrix form

$$K(CV + DW) = W$$

1.2.1.1   Standard eigenstructure assignment

Output feedback case: Standard eigenstructure assignment concerns the assignment of $q = p$ triples (λ_i, v_i, w_i) where p is the number of measurements, therefore, provided that $CV + DW$ is non-singular:

$$K = [w_1 \ldots w_p](C[v_1 \ldots v_p] + D[w_1 \ldots w_p])^{-1} \qquad (1.32)$$

or in matrix form

$$K = W(CV + DW)^{-1}$$

If $p < n$, the $n - p$ non-assigned poles are not troublesome if they correspond to weakly controllable (see page 99) or to fast enough dynamics (non-dominant modes). Otherwise Lemma 1.2.1 or an observer (see page 135) can be helpful.

State feedback case: The state feedback case is similar to the output feedback case, it suffices to replace p by n, C by I_n and D by $0_{p,m}$. So,

$$K = [w_1 \ldots w_n][v_1 \ldots v_n]^{-1} \qquad (1.33)$$

or in matrix form

$$K = WV^{-1}$$

1.2.1.2   Under-specified assignment

When the number (say $= q$) of triples (λ_i, v_i, w_i) that must be assigned is less than p (or n in the state feedback case), some degrees of freedom remain for choosing the feedback gain.

Case with no reference feedback. It seems natural to look for the gain of minimum norm which performs the considered assignment. The linear constraint induced by assignment is:

$$K(C[v_1 \ldots v_q] + D[w_1 \ldots w_q]) = [w_1 \ldots w_q] \tag{1.34}$$

Let us denote

$$Q = C[v_1 \ldots v_q] + D[w_1 \ldots w_q]$$

Equation (1.34) becomes $KQ = W$. The gain with minimum norm satisfying this constraint is

$$K = [w_1 \ldots w_q](Q^T Q)^{-1} Q^T \tag{1.35}$$

Reference feedback case. If there is a reference feedback gain K_0, it is natural to minimize $\|K - K_0\|$ when (1.34) is solved. For that purpose, the constraint $KQ = W$ can be written $(K - K_0)Q = W - K_0 Q$. The solution given in (1.35) becomes

$$K = K_0 + ([w_1 \ldots w_q] - K_0 Q)(Q^T Q)^{-1} Q^T \tag{1.36}$$

that minimizes $\|K - K_0\|$.

1.2.1.3   Feedback gain computed as a real matrix

In all cases, computing K that assigns a set of selected triples (v_i, w_i, λ_i) consists of solving a matrix equation of the form given by Equation (1.31). It remains to justify the fact that there exists a real feedback gain K for such assignments.

The key point for that comes from the assumption of page 15 that is: if λ_i is non-real, it is assumed that for some i', $\lambda_{i'} = \overline{\lambda}_i$ is also assigned and that $v_{i'} = \overline{v}_i$, $w_{i'} = \overline{w}_i$. For justifying the fact that this assumption is sufficient for the existence of a real gain, it suffices to consider the specific assignment of a pair of complex conjugate triples, the generalization being straightforward. So, it is assumed that (v_i, w_i, λ_i) and $(\overline{v}_i, \overline{w}_i, \overline{\lambda}_i)$ are simultaneously assigned. In view of (1.7),

$$K(C[v_i \ \overline{v}_i] + D[w_i \ \overline{w}_i]) = [w_i \ \overline{w}_i]$$

Multiplying this equation on the right by

$$\begin{bmatrix} 1 & -j \\ 1 & j \end{bmatrix}$$

an equivalent equation is obtained:

$$K(C[\Re(v_i)\ \Im(v_i)] + D[\Re(w_i)\ \Im(w_i)]) = [\Re(w_i)\ \Im(w_i)]$$

In this equation all coefficients are real numbers. Therefore, there exists a real solution to the eigenstructure assignment problem.

Note that, applying the same transformation to (1.30), we obtain the following real equivalent form:

$$A[\Re(v_i)\ \Im(v_i)] + B[\Re(w_i)\ \Im(w_i)] = [\Re(v_i)\ \Im(v_i)] \begin{bmatrix} \Re(\lambda_i) & \Im(\lambda_i) \\ -\Im(\lambda_i) & \Re(\lambda_i) \end{bmatrix}$$

1.2.1.4   Summary

Select $q \le p$ triples (λ_i, v_i, w_i) satisfying Equation (1.6). Depending on the design objectives, each eigenvector assignment can be subject to additional constraints which are

- Equation (1.22) for decoupling.

- Equations (1.24)-(1.26) vector projection case.

- Equations (1.27)-(1.29) minimum energy case.

Note that when λ_i is non-real, the triple $(\overline{\lambda}_i, \overline{v}_i, \overline{w}_i)$ must also belong to the set of assigned triples.

Then, compute the feedback gain by using (1.32) if $q = p$ or (1.36) if $q < p$.

Recommendation. *If there are no known relevant decoupling requirements, proceed by "continuity" as much as possible by assigning projections of eigenvectors and by shifting "slowly" the poles. If open-loop poles are well located re-assign them at the same location with the same eigenvector. This treatment must be reserved to dominant poles. During this "continuous motion" of dominant poles, analyze the motion of weakly controllable poles and stop when necessary. If there are more than p dominant poles to be treated consider a dynamic feedback (for example an observer which adds a number of observed signals equal to the number of missing measurements, see §1.2.8).*

Software

Related MATLAB-**function**: `fb_prop` and `eig_cstr`. The first function
is used for proportional feedback design. `eig_cstr` performs the same
computation but does not compute the gain, the output arguments of
this function are treated in a more complex way in order to compute
multi-model / dynamic feedback (see `fb_dyn` in §1.3). The assignment
options `key = 'z', 'n', 'v', 'p', 'm'` are defined on pages 22, 23
and 25. The use of `fb_prop` is illustrated below by three examples.

Example 1. Lateral control of an aircraft is the most popular example of
applications of eigenstructure assignment. Let us consider the linearized
model obtained using `rcamdata` (see page 291 for the definition of state,
input and output signals).

```
sys = rcamdata('lat',0,1);
```

The open-loop eigenvalues are $\{-0.23 \pm 0.60i, -1.30, -0.18\}$. We would
like to assign

$$\lambda_1 = -0.6 \pm 0.6i \,, \; v_1 = \begin{bmatrix} * \\ 0 \\ * \\ 0 \\ * \\ 0 \end{bmatrix} \; ; \; \lambda_2 = -0.6 \,, \; v_2 = \begin{bmatrix} 0 \\ * \\ 0 \\ * \\ 0 \\ * \end{bmatrix} \; ; \; \lambda_3 = -1.3 \,, \; v_3 = \begin{bmatrix} 0 \\ * \\ 0 \\ * \\ 0 \\ * \end{bmatrix}$$

in which "$*$" denotes non-constrained entries. This over-specified problem is solved automatically by least squares minimization by invoking
`fb_prop`. For that purpose, the constraint relative to v_1 can be written

$$\begin{bmatrix} 0 & 1 & 0 & 0 & 0 & 0 \\ 0 & 0 & 0 & 1 & 0 & 0 \\ 0 & 0 & 0 & 0 & 0 & 1 \end{bmatrix} v_1 = \begin{bmatrix} 0 \\ 0 \\ 0 \end{bmatrix} \tag{1.37}$$

So, we can use:

```
pol = [-0.6+0.6*i -0.6 -1.3];
key = ['nnn'];
CC = eye(6,6); DD = zeros(6,2);
def_pb = [ [2;4;6] [1;3;5] [1;3;5] ];
```

Note that the complex conjugate value $-0.6 - 0.6i$ is automatically assigned and that the matrix `def_pb` selects the entries of the identity
matrix `CC` (`CC(def_pb(:,1))`) is the left hand side matrix of Equation
(1.37)).

```
K = fb_prop(sys,0,pol,key,def_pb,CC,DD);
```

Example 2. Alternatively, the same problem can be treated using the assignment option `'z'` instead of `'n'`. In this case, it is not necessary to define the matrices CC and DD.

```
pol = [-0.6+0.6*i -0.6 -1.3];
key=['zzz'];
def_pb = [ [2;4;6] [1;3;5] [1;3;5] ];
K = fb_prop(sys,0,pol,key,def_pb);
```

Example 3. Instead of minimizing the size of some entries of right eigenvectors, it is also interesting to decouple some given outputs from some modes. Assume that the closed-loop mode at $-0.6\pm0.6i$ is expected to have no effect on the 3rd measurement signal (**sys.c** is denoted C) *i.e.*, `C(3,:)*v1 = 0` and that the closed-loop modes $\{-0.6, -1.3\}$ are expected to have no effect on the 2nd measurement *i.e.*, `C(2,:)*v2 = 0`, `C(2,:)*v3 = 0`. For that purpose:

```
pol = [-0.6+0.6*i -0.6 -1.3];
key = ['nnn'];
def_pb = [ 3 2 2 ];
```

Here `def_pb` selects the relevant row of the matrix **sys.c**.

```
K = fb_prop(sys,0,pol,key,def_pb);
```

Other examples: see pages 119, 131, 136 and the on-line **help2** message of **fb_prop**.

1.2.2   Pole assignment by output feedback

Most pole assignment techniques are special cases of the framework described in [Magni and Champetier, 1991]. When more than p poles are to be assigned, the most elementary technique of this reference is detailed in Lemma 1.2.1, it enables us to assign at most $m + p - 1$ poles (see [Champetier and Magni, 1991] for the theory).

Up to now, the vectors v_i and w_i were selected together. In the following lemma, it will be necessary to identify the vector w_{q_1+1} after v_{q_1+1} is selected. The eigenvectors that can be assigned as closed-loop eigenvectors relative to λ_{q_1+1} belong to the column span of the matrix $V(\lambda_{q_1+1})$ defined by

$$\begin{bmatrix} A - \lambda_{q_1+1}I & B \end{bmatrix} \begin{bmatrix} V(\lambda_{q_1+1}) \\ W(\lambda_{q_1+1}) \end{bmatrix} = 0$$

therefore, when v_{q_1+1} is chosen, there exists a vector ξ such that $v_{q_1+1} = V(\lambda_{q_1+1})\xi$. This vector being computed, the corresponding vector w_{q_1+1} is given by

$$w_{q_1+1} = W(\lambda_{q_1+1})\xi \tag{1.38}$$

LEMMA 1.2.1 *(D = 0) Let us consider q_1, triples (λ_i, v_i, w_i), $q_1 < p$ satisfying (1.6). Then consider an additional non-zero vector v_{q_1+1} satisfying*

$$v_{q_1+1} \in \text{Im } V(\lambda_{q_1+1}) \cap ((A - \beta_1 I)\text{Ker}C + \text{Im } [v_1 \ldots v_{q_1}]) \cap \ldots$$

$$\cap ((A - \beta_{q_2} I)\text{Ker}C + \text{Im } [v_1 \ldots v_{q_1}]) \quad (1.39)$$

Then K such that

$$KC[v_1 \ldots v_{q_1} \; v_{q_1+1}] = [w_1 \ldots w_{q_1} \; w_{q_1+1}] \quad (1.40)$$

assigns the $q_1 + 1$ triples (λ_i, v_i, w_i), $i = 1, \ldots, q_1 + 1$ and the additional q_2 eigenvalues $\beta_1, \ldots, \beta_{q_2}$.

Proof. See Appendix 4, page 265. The above result holds if $D = 0$ but if $D \neq 0$, apply the proposed technique ignoring D and then transform the obtained gain using (4.3).

The difficulty behind this formulation consists of finding the largest values of q_1 and q_2 for which the intersection in (1.39) does not reduce to zero. This lemma is often used with $q_1 = p - 1$ and $q_2 = m - 1$. In this case $q_1 + q_2 + 1 = m + p - 1$ poles are assigned. It comes out that

$$m + p > n \quad (1.41)$$

is a sufficient condition for assignment of all poles.

The summary of §1.2.1 describes a design procedure that is consistent with the formulation of Lemma 1.2.1. A simple adaptation has to be added, that is the $(q_1 + 1)$th eigenvector must belong to the intersection in Equation (1.39). The corresponding value of w_{q_1+1} can be computed from (1.38).

The result of Lemma 1.2.1 will be adapted to dynamic feedback in §1.2.6.

```
Software
```

Related MATLAB-**function:** fb_prop. The option relative to the $(q_1 + 1)$th vector is key = 'e'. For the q_1 first vectors use options 'z', 'n' (see page 22), 'p', 'v' (see page 23) or 'm' (see page 25).

Example. Assume that $n = 4$, $m = 2$, $p = 3$. One complex eigenvector associated with the assigned eigenvalue $\lambda_1 = -1 + j$ must be such that the first entry of the input direction is equal to zero. The assignment of an eigenvalue $\lambda_3 = -2$ must induce the assignment of an additional eigenvalue $\beta_1 = -5$. (The notations of Lemma 1.2.1 are used below.)

```
sys=rss(4,3,2);
lambda1 = -1+j;
lambda3 = -2;
beta1 = -5;
pol = [lambda1 lambda3];
key = ['z' 'e' ];
def_pb = [ 5 beta1 ];
```

Here, `def_pb(1,1)` = 5 means that the 5th entry of the vector (v_1, w_1) *i.e.*, the 1st one of w_1 is set to zero (option `'z'`) and `def_pb(1,2)` = `beta1` means that the assignment of `lambda3` induces the assignment of `beta1` as in Lemma 1.2.1 (option `'e'`).

```
K = fb_prop(sys,0,pol,key,def_pb);
```

1.2.3   Insensitive state feedback design

The feedback gain K is computed relative to the data available in the matrices (A, B, C, D). When these matrices are subject to large variations, the problem of preserving stability and performance is a problem of "*robustness*". Here, in this paragraph, small variations are considered, so we will speak of "*insensitivity*" instead of "robustness". More precisely this paragraph deals with the insensitivity of the closed-loop eigenvalues.

First, closed-loop eigenvalue sensitivity is translated in terms of eigenstructure. Consider the closed-loop system $\dot{x} = \hat{A}x$ in which $\hat{A}{=}A + B((I - KD)^{-1}K)C$. Let λ_i be an eigenvalue of the matrix $\hat{A}$ and v_i, u_i be the corresponding right and left eigenvectors. The matrix $\hat{A}$ is subject to parameter variations modeled by $\Delta\hat{A}$.

LEMMA 1.2.2 *Assuming that u_i and v_i are normalized such that $u_i v_i = 1$, the formula*

$$\Delta\lambda_i = u_i \Delta\hat{A} \, v_i \tag{1.42}$$

gives the first order variation of the ith eigenvalue of the matrix $\hat{A}$.

Two particular cases of this result are useful:

- *Non-structured sensitivity :* $\Delta\hat{A}$ is a small variation which is not modeled:

$$\Delta\lambda_i \leq \|u_i\|\|v_i\| \times M \tag{1.43}$$

  in which M is some upper bound of the variations $\Delta\hat{A}$.

- *Structured sensitivity :* $\Delta \widehat{A}$ is induced by the variations of A, B, C, D which are respectively denoted $\Delta A, \Delta B, \Delta C, \Delta D$:

$$\Delta \lambda_i = u_i \Delta A v_i + u_i \Delta B w_i + t_i \Delta C v_i + t_i \Delta D w_i \qquad (1.44)$$

Proof. See Appendix 4, page 255.

Considering Equation (1.43), eigenstructure assignment for minimizing the sensitivity of closed-loop eigenvalues can be viewed as a problem of minimizing the sum of the products $\|u_i\|\|v_i\|$:

$$J = \sum_{i=1}^{n} \|u_i\|\|v_i\| \qquad (1.45)$$

Geometric interpretation of eigenvalue sensitivity. A pair of real eigenvalues / eigenvectors is considered in this discussion. As $u_i v_i = 1$, $\|u_i\|\|v_i\|$ is one over the cosine of the angle between the vectors u_i^* and v_i. Furthermore, u_i^* is orthogonal to $\text{Im}[v_1, \ldots, v_{i-1}, v_{i+1} \ldots, v_n]$ (remember that $UV = I$), so, $\|u_i\|\|v_i\|$ is the sine between v_i and the other right eigenvectors. It turns out that

Maximizing the angles between the right eigenvectors minimizes in some sense the sensitivity of closed-loop eigenvalues.

Unlike in §1.1.6 to 1.1.8, for reducing sensitivity, the assignment of a single eigenvector cannot be considered independently of other assignments (the criterion (1.45) depends on all eigenvectors). However, if a controller is given and if it is expected to reduce the sensitivity of a single closed-loop eigenvalue (the ith), in view of the above discussion, the ith right eigenvector must be changed so that ideally it becomes equal to u_i^*. In a more realistic way, it is the orthogonal projection of u_i^* that must be considered, therefore, from (1.24) to (1.26) (page 23),

$$[\, A - \lambda_i I \quad B \,] \begin{bmatrix} V(\lambda_i) \\ W(\lambda_i) \end{bmatrix} = 0$$

$$\eta_i = (V^*(\lambda_i)V(\lambda_i))^{-1}V^*(\lambda_i)u_i^* \qquad (1.46)$$

and

$$v_i = V(\lambda_i)\eta_i \qquad (1.47)$$

$$w_i = W(\lambda_i)\eta_i \qquad (1.48)$$

Now, the technique presented above is used in an algorithm that minimizes the sensitivity of all closed-loop poles. This algorithm is heuristic (no convergence proof) but it has shown to be very efficient in many applications. It consists of applying the assignment of (1.46)-(1.48) in turn to all right eigenvectors. For simplifying the description of the algorithm, it is first assumed that all closed-loop eigenvalues are real.

Algorithm.

1 - Initialization: select the closed-loop poles and compute an initial pole assignment by state feedback. This step consists of applying (1.33) with a random selection of closed-loop eigenvectors. Several random trials can be considered, the best one being chosen before going to Step 2. Set $i = 1$ and compute the matrices $V = [v_1 \ldots v_n]$ and $W = [w_1 \ldots w_n]$.

2 - All the column vectors of the matrix V, except the ith one are fixed. Compute u_i the orthogonal to $[v_1 \ldots v_{i-1}, v_{i+1}, \ldots v_n]$ and then, the optimal vector $\widehat{v}_i$ as in (1.47). The corresponding input direction $\widehat{w}_i$ is computed using (1.48). Replace v_i by $\widehat{v}_i$ in V and w_i by $\widehat{w}_i$ in W.

3 - Compute the criterion (1.45). If the criterion does not reduce any more or oscillates, *stop the algorithm* and compute the "optimal" state feedback. The gain can be computed using (1.33).

4 - Set i to $i + 1$ if $i < n$ or to 1 if $i = n$ and go to Step 2.

Usually, this algorithm must be completed by standard optimization. The advantage of using this algorithm before standard optimization is that, in almost all cases, optimization will not lead to a very bad local minimum.

Extension to the output feedback case. This technique can be adapted to output feedback design. In this case, only p right eigenvectors are treated. So, u_i cannot be computed as the orthogonal to $n - 1$ vectors as only $p - 1$ vectors v_i are known. Therefore, for computing u_i, it is necessary to compute the feedback gain (using (1.32) instead of (1.33)) and then the matrix of left eigenvectors U from which u_i can be extracted. Such computation is required each times one of the p vectors v_i is updated. This technique might be troublesome if the $n - p$ ignored poles are somewhat controllable because reducing the sensitivity tends to spread poles over a larger area. Often, this technique results in instability. It is possible to avoid instability by considering additional constraints as shown on page 92.

Adaptation to complex eigenvalues. In the complex case, the optimal vector $\widehat{v}_i$ is computed using the same formula than in the real case. The difference lies in the fact that both vectors v_i and $\overline{v}_i$ are updated respectively by $\widehat{v}_i$ and $\widehat{\overline{v}}_i$ (the algorithm must be adapted so that the update of $\overline{v}_i$ is not treated).

Software

Related MATLAB-**function:** sfb_ins. The dual treatment for observers will be presented later (function ob_ins). The function sfb_ins is better used for state feedback design, however, an option permits the user to treat the output feedback case (option opt = 'o').

Example: See page 122.

1.2.4 Non-interactive control design

Output feedback for approximate non-interactive control design (or decoupling) by eigenstructure assignment is presented here. The theoretical justification for *exact* decoupling by *state feedback* can be found in Appendix 4.5, page 268. Approximate decoupling is discussed at the end of §4.5. A comparison of the output feedback approach with alternative approaches (using an observer or a dynamic feedforward) can be found in §2.3.

Problem statement. The following system is considered

$$\begin{aligned} \dot{x} &= & Ax &+ & Bu \\ y &= & Cx &+ & Du \\ z &= & Ex & & \end{aligned} \tag{1.49}$$

in which z is a vector of regulated outputs. We look for a feedback law

$$u = Ky + Hz_R \tag{1.50}$$

such that each reference input z_{Ri} affects z_i but has no cross effect on z_k for $k \neq i$.

We shall first consider the steady state solution of this problem. The more complex transient case will be treated in a second step.

1.2.4.1 Steady state decoupling by constant feedforward design

H is computed so that the steady state gain between the reference input vector z_R and the regulated output vector z is the identity matrix. Let us consider the system of Equation (1.49) in which temporarily F is nonzero, $z = Ex + Fu$. The feedback gain K is assumed to be already

designed. It is assumed that the number of rows of $[E \ F]$ is lower or
equal to the number of columns of B.

The computation of H is quite simple. First, the control loop is closed,
the above system becomes:

$$\dot{x} = \widehat{A}x + \widehat{B}Hz_R$$
$$z = \widehat{E}x + \widehat{F}Hz_R$$

for some matrices $\widehat{A}$, $\widehat{B}$, $\widehat{E}$ and $\widehat{F}$. Then the steady state gain between
z_R and z is $-\widehat{E}\widehat{A}^{-1}\widehat{B}H + \widehat{F}H$, therefore

$$H = (-\widehat{E}\widehat{A}^{-1}\widehat{B} + \widehat{F})^{-1}$$

In the rectangular case, the minimum norm solution can be considered.
If the feedback K is dynamic, there is no difficulty for generalizing this
result.

1.2.4.2   Eigenstructure assignment constraints for transient decoupling

The modal point of view for non-interactive control design is illustrated in Figure 1.3. We use for that purpose a simple example in which
only two reference inputs and regulated outputs are considered. The
generalization will be given in a summary. Used notations:

$$z_1 = E_1 x$$
$$z_2 = E_2 x$$

The matrix H is denoted

$$H = [\ H_1 \quad H_2 \ ]$$

There are four classes of closed-loop modes appearing in this Figure 1.3.

- *First class of modes:* Modes associated with the output z_1,
  $\leftrightarrow$ modes which are "not seen from" z_2, i.e., $E_2 v_i = 0$
  $\leftrightarrow$ modes which are not affected by z_{R2}, i.e., $u_i B H_2 = 0$

- *Second class:* Modes associated with the output z_2,
  $\leftrightarrow$ modes which are "not seen from" z_1, i.e., $E_1 v_i = 0$
  $\leftrightarrow$ modes which are not affected by z_{R1}, i.e., $u_i B H_1 = 0$

- *Third class.* The third class corresponds to invariant zeros (see below)
  between the inputs and the regulated outputs. For these modes, the
  right eigenvectors must be chosen such that $E_1 v_i = 0$ and $E_2 v_i = 0$.
  Note that these modes are not visible from both regulated outputs,

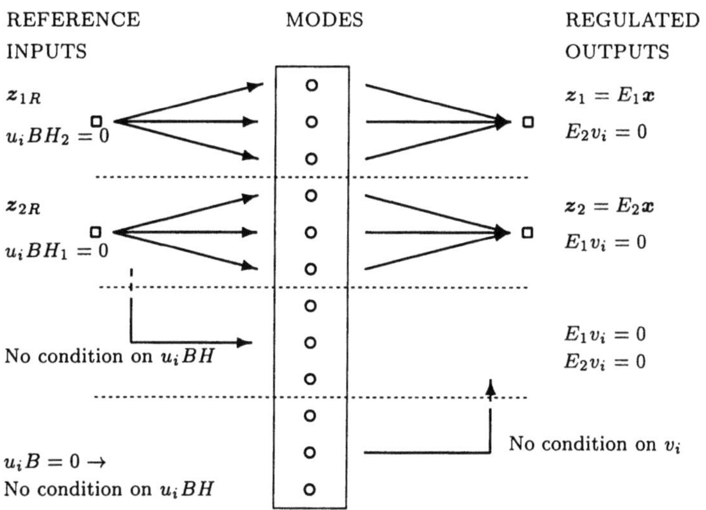

Figure 1.3.   Modal illustration of non-interactivity

so there is no condition relative to the transmissions from reference inputs to the modes.

- *Fourth class.* The fourth class concerns non-controllable eigenvalues. Non controllability is equivalent to $u_i B = 0$ (u_i left eigenvectors of A, see page 247). So, there is no transmission from reference inputs to these modes (no additional eigenstructure assignment constraint).

The third class corresponds to invariants zeros of the triple (A, B, E). The zeros are the complex numbers $s = z_i$ at which the rank of the system matrix

$$\begin{bmatrix} A - sI & B \\ E & 0 \end{bmatrix}$$

becomes less than the normal rank[4]. So, if z_i is a zero, there exists a non-zero vector v_i such that

$$\begin{bmatrix} A - z_i I & B \\ E & 0 \end{bmatrix} \begin{bmatrix} v_i \\ w_i \end{bmatrix} = 0$$

1.2.4.3   Discussion relative to the feedforward constraints

The assignment of the right eigenvectors as above fixes the feedback gain. Therefore there is no remaining freedom for the left eigenvectors.

For taking into account the transmissions from the reference inputs towards the modes some constraints relative to the feedforward gain H must be considered. From the above discussion, these constraints are of the form

$$u_i B H_k = 0$$

As H_k is a column vector and $m \geq 2$, H_k can be computed by fixing one of its entries and by solving the resulting equations for non-fixed entries. Least squares can be used if there are too many equations.

In fact, as explained in Appendix 4.5, if exact decoupling is possible, the matrix H that satisfies all relevant constraints of the form $u_i B H_k = 0$ is the one defined in §1.2.4.1. It is this form that is adopted in the sequel. However, in some cases, the former form can be better[5].

1.2.4.4   Summary

This summary gives an adaptation of the exact design procedure that is described in §4.5. It is assumed that there are $q \leq m$ regulated outputs and a number of measurements at least equal to the number of dominant modes.

Algorithm.

1 - Compute the zeros of the triple (A, B, E). Let n_z denote the number of zeros (it is assumed that all zeros are in the left half plane). For each zero (denoted z_i) solve

$$\begin{bmatrix} A - \lambda_{zi} I & B \end{bmatrix} \begin{bmatrix} v_{zi} \\ w_{zi} \end{bmatrix} \text{ and } E v_{zi} = 0$$

2 - Chose q integers m_i such that

$$n = n_z + m_1 + \ldots + m_q$$

For $i = 1, \ldots, q$ select a set of m_i self conjugate closed-loop eigenvalues and solve for each selected eigenvalue λ_{ik}

$$\begin{bmatrix} A - \lambda_{ik} I & B \end{bmatrix} \begin{bmatrix} v_{ik} \\ w_{ik} \end{bmatrix} \text{ and } E_j v_{ik} = 0 \ \forall j \neq i$$

3 - The matrices V (resp. W) being defined from the above vectors v_{zi}, v_{ik} (resp. w_{zi}, w_{ik}), compute K such that

$$K(CV + DW) = W$$

4 - Compute the minimum norm matrix H such that

$$(-\hat{E}\hat{A}^{-1}\hat{B})H = I$$

in which the "hat" denotes matrices after closure of the feedback loop.

Comments. The integer numbers m_i look like the numbers $n_i + 1$ defined in the appendix (page 268). But these numbers are valid in the state feedback case. Here, output feedback is used, so, some poles are not explicitly assigned. In this case, the interest of computing the integers n_i becomes doubtful. But as there is usually a limited number of possibilities for selecting the numbers m_i, trials and errors may replace mathematical analysis.

However, zero analysis, as in the state feedback case, remains essential (see the software part below).

Software

Related MATLAB-**function**: fb_prop and ff_stat. The first function is used for decoupling by right eigenvectors assignment, the second one is used for steady state decoupling.

Example 1. This example concerns the RCAM model (see page 290). The regulated outputs that are considered for decoupling are respectively the 2nd and 3rd measurements (more realistic decoupling is considered in §2.3). Four measurements are available for four dominant poles to be assigned.

```
sys=rcamdata('lat',0,1);
```

The invariant zeros from inputs to regulated outputs $z_1 = y_2$ and $z_2 = y_3$ are computed as follows

```
tzero([0 1 0 0;0 0 1 0]*sys)
```

An alternative way consists of computing the almost zeros defined in §1.5.3 as follows [x1,x2]=azer([0 1 0 0;0 0 1 0]*sys,10,1e-6). Two zeros are found: 0 and -0.1397. Eigenstructure assignment is performed considering $m_1 = m_2 = 1$, so one pole (-0.4) is associated with $z_1 = y_2$ and one pole (-0.5) is associated with $z_2 = y_3$.

```
p_n =[-0.4 -0.5 -0.1397 0];
K1 = fb_prop(sys,0,p_n,'n',[3 2 2 2;0 0 3 3]);
```

Finally the feedforward gain is computed as follows

```
H1 = ff_stat(sys,K1,[2 3]);
```

Example 2. The following control design is similar to the previous one but one zero (-0.1397) is ignored. A pair of conjugate poles $-0.4 + 0.4j$ is associated with z_1.

```
sys=rcamdata('lat',0,1);
p_n=[-0.4+0.4*i -0.5 0];
K1 = fb_prop(sys,0,p_n,'n',[3 2 2;0 0 3]);
H1 = ff_stat(sys,K1,[2 3]);
```

Figure 1.4 shows the effect of ignoring one of the zeros. Cross coupling (from input 2 to regulated output 1) becomes 100 times more important in the second case (zero ignored, right simulation) than in the first one (left simulation).

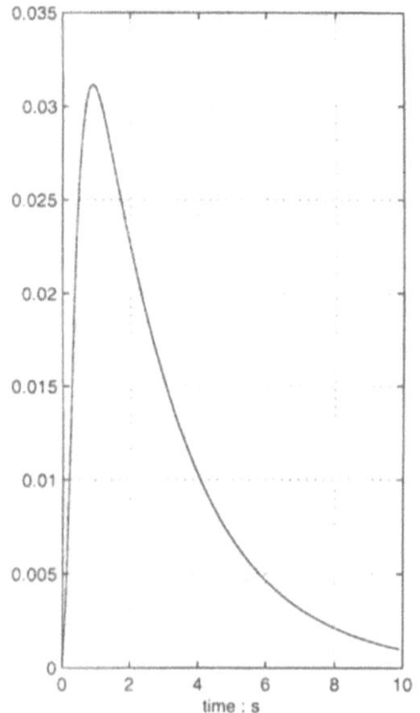
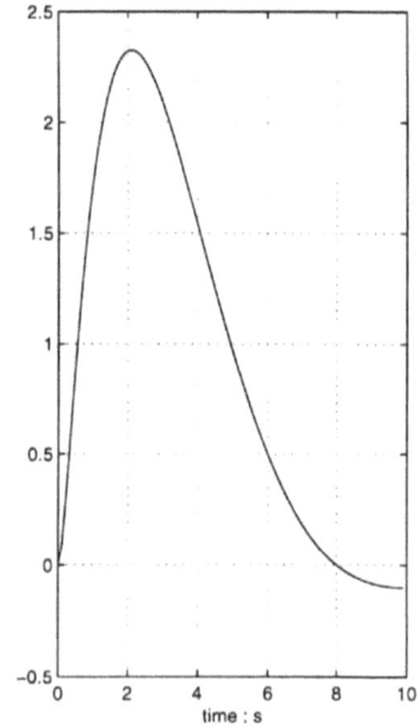

Figure 1.4.   Comparison of transient cross-coupling (one zero ignored on the right)

Other examples: see §2.3 and concerning `ff_stat`, pages 44, 130 and the on-line `help2` message of `ff_stat`.

1.2.5   Dynamic feedforward

Right eigenvectors are assigned by feedback, but it is possible to replace all or a part of this assignment by the use of a dynamic feedforward. The underlying idea for that consists of a generalization of pole / zero cancellation from the SISO to the MIMO case. For SISO systems, a feedforward gain cancels a pole by introducing a zero at the same location (see Figure 1.5). A new pole is also introduced. In fact the cancelled pole becomes non-controllable and from the input / output point of view, it seems that it has been shifted to the new pole location. In the MIMO case, modes are characterized by a pole *and an eigenvector*. The proposed MIMO generalization takes this fact into account.

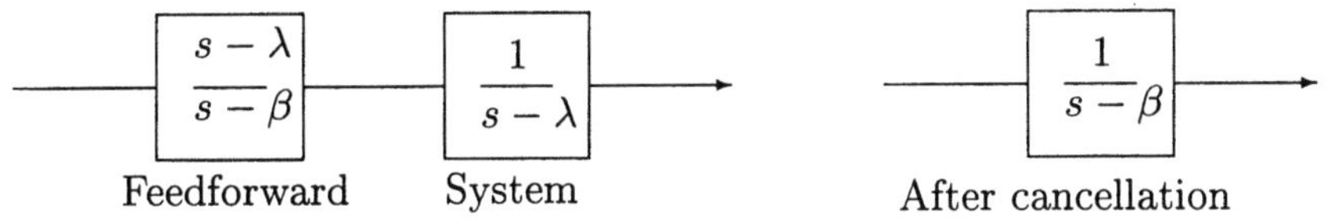

Figure 1.5.   SISO cancellation.

Eigenstructure "assignment" by means of dynamic feedforward instead of feedback can be justified for example when we do not want to devote all the feedback design degrees of freedom to decoupling. Feedback may be devoted to robustness while decoupling is made by feedforward.

Let us consider the feedforward state space model (A_f, B_f, C_f, D_f). Without loss of generality we can assume that $D_f = I$.

$$\begin{aligned}
\dot{x}_f &= A_f x_f + B_f v \\
y_f &= C_f x_f + I v
\end{aligned} \qquad (1.51)$$

If the matrix A is not stable, a stabilizing output feedback gain K_0 must be first designed. Hereafter, A denotes a stable matrix. The output of System (1.51) is connected to the input of System (1.1) $(u = y_f)$. The global systems can be modeled as:

$$\begin{bmatrix} \dot{x} \\ \dot{x}_f \end{bmatrix} = \begin{bmatrix} A & BC_f \\ 0 & A_f \end{bmatrix} \begin{bmatrix} x \\ x_f \end{bmatrix} + \begin{bmatrix} B \\ B_f \end{bmatrix} v \qquad (1.52)$$

We are looking for (A_f, B_f, C_f, I) so that

- q eigenvalues of A denoted $\{\lambda_1 \ldots \lambda_q\}$ are made uncontrollable.

- q new arbitrarily chosen eigenvalues $\{\beta_1 \ldots \beta_q\}$ are introduced.

- The eigenvectors corresponding to these new eigenvalues are assigned with the same degrees a freedom as in the state feedback case (for example satisfying $Ev_i + Fw_i = 0$).

For that purpose:

LEMMA 1.2.3 *The dynamic feedforward system* (A_f, B_f, C_f, I) *defined below cancels* $\lambda_1, \ldots, \lambda_q$ *and replace these eigenvalues by* $\beta_1 \ldots \beta_q$ *($\beta_i \notin \sigma(A)$) with associated eigenvectors* (v_i, v_{ci}) *such that* $Ev_i + Fw_i = 0$.
1 - $B_f = -U_q B$ *where* U_q *is the matrix of the* q *left eigenvectors of* A

corresponding to $\lambda_1 \ldots \lambda_q$.

2 - $C_f = [w_1 \ldots w_q][v_{c1} \ldots v_{cq}]^{-1}$ *where*

$$\begin{bmatrix} A - \beta_i I & 0 & B \\ 0 & \Lambda_q - \beta_i I & B_f \\ E & 0 & F \end{bmatrix} \begin{bmatrix} v_i \\ v_{ci} \\ w_i \end{bmatrix} = 0$$

3 - $A_f = \Lambda_q + B_f C_f$.

Proof. See Appendix 4, page 277.

When the matrices A_f, B_f and C_f are obtained after having followed the design procedure stated in Lemma 1.2.3, it might be relevant to multiply $D_f = I$ and B_f by the matrix $(I - C_f A_f^{-1} B_f)^{-1}$ in order to normalize the steady state gain.

Software

Related MATLAB-**function**: `ff_assgn`. This function performs the design procedure detailed in Lemma 1.2.3 with the aforementioned normalization. In addition, this function permits the user to define approximate (optimal) cancellation when the poles to be cancelled are not equal to the zeros.

Example. A random stable system is considered.

```
rand('seed',0); randn('seed',0);
sys = rss(4,4,2);
```

A feedback gain K and the equivalent dynamic feedforward gain `ff` are designed (`clean_ev` remove complex conjugate values):

```
K = fb_prop(sys,0,[-1 -2 -3+3*j],'z',[1 2 3]);
```

Feedforward computation

```
zer = clean_ev(eig(sys));
ff = ff_assgn(sys,0,zer,zer,[-1 -2 -3+3*j],'z',[1 2 3]);
```

Before comparison of feedback and feedforward controllers, the steady state gains between inputs and outputs must be rendered equal in order to have similar behavior in steady state (and, by the way, during the transient phase that is related to eigenvectors).

```
dc1 = dcgain(feedback(sys,K,1));
dc2 = dcgain(sys);
H = inv(dc2'*dc2)*dc2'*dc1
```

The last code line solves the equation `dc1 = dc2*H`. Finally, it can be checked that the simulations obtained below are identical.

```
figure
lsim(feedback(sys,K,1),ones(101,2),0:0.1:10)
figure
lsim(sys*ff*H,ones(101,2),0:0.1:10)
```

It is also possible to assign a subset of eigenvectors by feedback and the complementary subset by feedforward.

Other examples: see page 134 and the on-line `help2` message of `ff_assgn`.

1.2.6 Dynamic extension for output feedback design

Dynamic feedback design will be considered using observers in §1.2.8-1.2.9 and using a transfer function matrix approach in §1.3. Here is presented a much more elementary technique that can be used in a straightforward way for improving the applicability of the techniques proposed in §1.2.1 to 1.2.3.

Definition of a dynamic extension. With dynamic extension of order n_c, System (1.1) becomes :

$$\left[\begin{array}{c} \dot{x} \\ \dot{x}_c \end{array} \right] = \left[\begin{array}{cc} A & 0 \\ 0 & \Omega \end{array} \right] \left[\begin{array}{c} x \\ x_c \end{array} \right] + \left[\begin{array}{cc} B & 0 \\ 0 & I_{n_c} \end{array} \right] \left[\begin{array}{c} u \\ u_c \end{array} \right]$$

$$\left[\begin{array}{c} y \\ y_c \end{array} \right] = \left[\begin{array}{cc} C & 0 \\ 0 & I_{n_c} \end{array} \right] \left[\begin{array}{c} x \\ x_c \end{array} \right] + \left[\begin{array}{cc} D & 0 \\ 0 & 0 \end{array} \right] \left[\begin{array}{c} u \\ u_c \end{array} \right] \qquad (1.53)$$

In which Ω is any real $n_c \times n_c$ matrix (usually $\Omega = 0_{n_c,n_c}$).

The benefits of using a dynamic extension are stated in the following lemma. Briefly, a dynamic gain can be viewed as a proportional gain relative to the extended system, by the way, the techniques presented in §1.2.1 to 1.2.3 have an immediate dynamic counterpart.

LEMMA 1.2.4 *The following feedback gains are equivalent:*

- $G(s)$ *applied to System (1.1), where*

$$G(s) = K_{11} + K_{12}(sI - K_{22} - \Omega)^{-1}K_{21} \qquad (1.54)$$

- $\overline{K}$ *applied to System (1.53), where*

$$\overline{K} = \left[\begin{array}{cc} K_{11} & K_{12} \\ K_{21} & K_{22} \end{array} \right] \qquad (1.55)$$

Proof. See Appendix 4, page 249.

Using the dynamic extension approach for dynamic feedback design is useful in view of Lemma 1.2.1: The numbers of states, inputs, outputs become respectively $(n + n_c, m + n_c, p + n_c)$ so that the condition $m + p > n$ (see (1.41)) becomes $m + p + n_c > n$. Therefore all eigenvalues can be assigned with a dynamic extension of order n_c at least equal to $n - (m + p) + 1$.

This dynamic extension will also be used for tuning a dynamic feedback. For that purpose, consider a dynamic feedback as in (1.54),

- transform it into the form of (1.55),

- tune this constant gain as proposed in §1.4 (considering System (1.53)),

- go back to the dynamic form of (1.54).

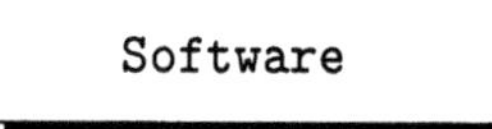

Related MATLAB-**functions**: add_dyn, sta2dyn and dyn2sta.

- add_dyn: transforms a system (A, B, C, D) into the system of Equation (1.53).

- dyn2sta: transforms a dynamic gain to K_{11}, K_{12}, K_{21}, K_{22} (taking into account the matrix Ω).

- sta2dyn: converse of dyn2sta.

Example: pole assignment. This example illustrates the use of the proposed dynamic extension in order to improve the potentialities for pole assignment of Lemma 1.2.1. Consider a system with 5 states, 2 inputs and 2 outputs. In order to assign all the eigenvalues a dynamic extension of order $n - (m + p) + 1 = 2$ must be considered.

```
sys1 = rss(5,2,2);
dyn=[-5;-10];
sys2 = add_dyn(sys1,dyn);
```

Computation of a proportional feedback. Let $\{\lambda_1 = -1 + j,\ \lambda_2 = -1 - j,\ \lambda_3 = -1,\ \lambda_4 = -3,\ \beta_1 = -2 + 2j,\ \beta_2 = -2 - 2,\ \beta_3 = -2\}$. (Zeros in def_pb in front of 'z' induce random eigenvector assignments.)

```
pol =   [-1+j      -1      -3 ];
key =   [  'z'     'z'     'e'];
def_pb = [ 0    0    -2+2*j;  ...
               0    0     -2 ];
Kd  = fb_prop(sys2,0,pol,key,def_pb);
```

It can be checked that the extended system has in closed-loop (constant gain Kd) the same eigenvalues as the original system with the dynamic feedback (fbdyn = sta2dyn(Kd,dyn)), in other words,

    eig_fb(sys2,Kd)
    eig_fb(sys1,fbdyn)

lead to the same result.

Other examples: See page 124 and the on-line help message of **dyn2sta**.

1.2.7   Observer definition

A modal approach to observer design is proposed in [Magni and Mouyon, 1991] and [Magni and Mouyon, 1994], see also [Magni, 1996]. This modal approach is based on "elementary observers" *i.e.*, observers of scalar signals. Combining elementary observers will permit us to define more general observers.

1.2.7.1   Elementary observers

LEMMA 1.2.5 *The system defined by (see Figure 1.6):*

$$\dot{\hat{z}} = \pi_i \hat{z} - t_i \boldsymbol{y} + u_i B \boldsymbol{u} + t_i D \boldsymbol{u} \tag{1.56}$$

where u_i, t_i, π_i *satisfy :*

$$u_i A + t_i C = \pi_i u_i \tag{1.57}$$

is an observer of the scalar signal $\boldsymbol{z} = u_i \boldsymbol{x}$. *The dynamics of the observation is given by* π_i *(i.e.,* $\varepsilon = \hat{z} - \boldsymbol{z}$ *satisfies* $\dot{\varepsilon} = \pi_i \varepsilon$*).*

Proof. See Appendix 4, page 278.

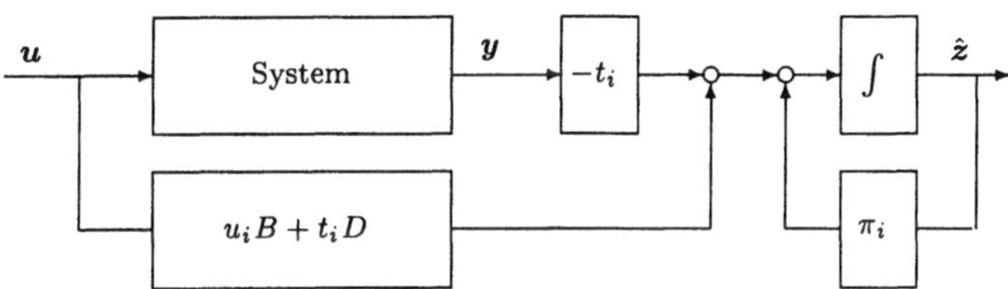

Figure 1.6.   Elementary observer of $\boldsymbol{z} = u_i \boldsymbol{x}$.

This lemma states that a linear combinations of states $u_i \boldsymbol{x}$ can be estimated by a one-dimensional observer provided that the vector u_i

satisfies (1.57) for some vector t_i and some complex number π_i. This equation can be written

$$[u_i \quad t_i] \begin{bmatrix} A - \pi_i I \\ C \end{bmatrix} = 0 \tag{1.58}$$

that shown some duality between elementary observer design and right eigenstructure assignment (see (1.6))[6].

1.2.7.2 General observers

If elementary observers are used in parallel, say a number equal to n_c, we can represent the overall observer as in Figure 1.6 but replacing u_i, t_i and π_i by matrix notations U, T and Π where

$$U = \begin{bmatrix} u_1 \\ \vdots \\ u_{n_c} \end{bmatrix} \; ; \; T = \begin{bmatrix} t_1 \\ \vdots \\ t_{n_c} \end{bmatrix} \; ; \; \Pi = \mathrm{Diag}\{\pi_1 \ldots \pi_{n_c}\} \tag{1.59}$$

where each triple (π_i, u_i, t_i) satisfy (1.56). These n_c equations can be written globally as :

$$UA + TC = \Pi U \tag{1.60}$$

Here, z becomes a n_c-dimensional vector. The observer equation becomes:

$$\dot{\hat{z}} = \Pi\hat{z} - Ty + UBu + TDu \tag{1.61}$$

in which $\hat{z}$ is an estimate of

$$z = Ux \tag{1.62}$$

Note that when π_i is non-real, the triple $(\overline{u}_i, \overline{t}_i, \overline{\pi}_i)$ must also be considered so that a simple transformation (see below, §1.2.7.3) permits us to manipulate matrices U, T and Π with real entries.

The following lemma states that the transfer function matrix between the input and the estimate $\hat{z}$ is equal to the one between the input and the corresponding estimated signal z.

LEMMA 1.2.6 *The transfer function matrix between u and $\hat{z}$ (see System (1.61)) is equal to $U(sI - A)^{-1}B$, which is also the transfer function matrix between u and z. The eigenvalues of Π are uncontrollable.*

Proof. See Appendix 4, page 281.

From the estimated variables $\hat{z}$, the inputs u and the measurements y, the set of signals η that can be estimated is described by three real matrices Q_u, Q_y and Q_z (with appropriate dimension):

$$\eta = \begin{bmatrix} Q_u & Q_y & Q_z \end{bmatrix} \begin{bmatrix} u \\ y \\ z \end{bmatrix} \qquad \text{estimate:} \quad \hat{\eta} = \begin{bmatrix} Q_u & Q_y & Q_z \end{bmatrix} \begin{bmatrix} u \\ y \\ \hat{z} \end{bmatrix}$$

This structure is depicted in Figure 1.7 (in which $Q_u = 0$). Note that the matrix Q_u is only used for subtracting the system direct transmission (for example, see Equation (1.65) below).

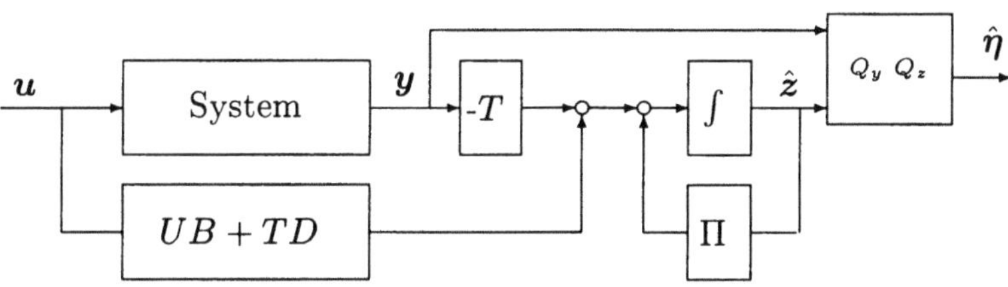

Figure 1.7. General observer ($Q_u = 0$ is omitted).

1.2.7.3 Observer with real coefficients

In order to manipulate real matrices, it is suggested to use a trick similar to the one of page 30. When an elementary observer (π_i, u_i, t_i) is designed with π_i non-real, it is understood that the conjugate elementary observer is also considered. Both equations of Lemma 1.2.5 can be written together

$$\begin{bmatrix} \dot{\hat{z}}_i \\ \dot{\bar{\hat{z}}}_i \end{bmatrix} = \begin{bmatrix} \pi_i & 0 \\ 0 & \bar{\pi}_i \end{bmatrix} \begin{bmatrix} \hat{z}_i \\ \bar{\hat{z}}_i \end{bmatrix} - \begin{bmatrix} t_i \\ \bar{t}_i \end{bmatrix} y + \begin{bmatrix} u_i & t_i \\ \bar{u}_i & \bar{t}_i \end{bmatrix} \begin{bmatrix} B \\ D \end{bmatrix} u \qquad (1.63)$$

and

$$\begin{bmatrix} u_i \\ \bar{u}_i \end{bmatrix} A + \begin{bmatrix} t_i \\ \bar{t}_i \end{bmatrix} C = \begin{bmatrix} \pi_i & 0 \\ 0 & \bar{\pi}_i \end{bmatrix} \begin{bmatrix} u_i \\ \bar{u}_i \end{bmatrix} \qquad (1.64)$$

Consider the following matrix Q

$$Q = \begin{bmatrix} 1 & 1 \\ -j & j \end{bmatrix}$$

Applying the change of basis defined by this matrix to Equations (1.63) and (1.64) (left multiplication), leads to equations in which all imaginary parts have disappeared. Even if this transformation is not always

mentioned in this book, it should be understood that it is applied to all
considered banks of observers.

1.2.7.4   State observers

State observers are observers defined by three matrices Q_u, Q_y and
Q_z satisfying

$$x = [\, Q_u \quad Q_y \quad Q_z \,] \begin{bmatrix} u \\ y \\ z \end{bmatrix}$$

that is equivalent to (using (1.62) and (1.1))

$$\begin{cases} Q_u = -Q_y D \\ I = [\, Q_y \quad Q_z \,] \begin{bmatrix} C \\ U \end{bmatrix} \end{cases} \tag{1.65}$$

If the number of rows of U is larger than $n - p$, generically it is possible
to find Q_y and Q_z as above.

The *minimum order state observer* corresponds to U with $n - p$ rows
such that

$$U \in \mathbb{R}^{(n-p) \times n}$$

$$Q_u = -DQ_y \quad \text{and} \quad [\, Q_y \quad Q_z \,] = \begin{bmatrix} C \\ U \end{bmatrix}^{-1} \tag{1.66}$$

The invertibility of the above matrix is also a generic property. For
notational convenience, Q_u will be ignored except in §1.2.9.

1.2.7.5   State observer with Kalman filter structure

The most popular observer structure is depicted in Figure 1.8. For
System (1.1), this structure corresponds to

$$\begin{aligned} \dot{\hat{x}} &= A\hat{x} + Bu + K_o(\hat{y} - y) \\ \hat{y} &= C\hat{x} + Du \end{aligned} \tag{1.67}$$

in which K_o is the observer gain. The gain K_o is designed like a state
feedback but in the dual way (see (1.33)). Therefore, the procedure is as
follows:

- select n triples (π_i, u_i, t_i) satisfying (1.58)

- considering the matrices U and T as defined in (1.59), compute $K_o = U^{-1}T$

LEMMA 1.2.7 *The state observer with Kalman filter structure in which the gain K_o is computed as above is equivalent to the observer of Figure 1.7 in which*

$$Q_y = 0 \quad and \quad Q_z = U^{-1} \tag{1.68}$$

The vector $\hat{x}$ in this figure denotes an estimate of the state vector.

Proof. See Appendix 4, page 280.

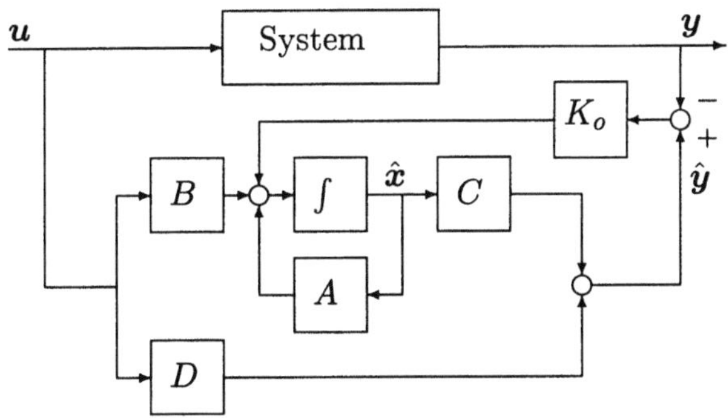

Figure 1.8.   Observer with Kalman filter structure

1.2.7.6   Elementary observer design by projection

We have defined general observers as combinations of elementary observers. Therefore, it suffices to discuss elementary observer design. Such design appears to be dual to right eigenvector assignment. We will dualize first the projection ideas of §1.1.7 and then, the decoupling ideas of §1.1.6.

Assume that λ_i is an open-loop eigenvalue which is well located (well damped, fast enough). Under similar conditions, in the feedback case (see page 30), we proposed to preserve the open-loop eigenstructure. By duality, the elementary observer which corresponds to this "assignment" is characterized by the open-loop triple $(\lambda_{0i}, u_{0i}, t_{0i})$ in which u_{i0} is a left eigenvector of A. This triple satisfies (1.57)-(1.58) for $t_{0i} = 0$ because

$$[u_{i0} \quad 0] \begin{bmatrix} A - \pi_i I \\ C \end{bmatrix} = 0$$

The corresponding observer is as illustrated in Figure 1.9. It turns out that the measurement vector y is not used for the observation of $z_i = u_{i0}x$. From our experience, this choice (or the closest one based on

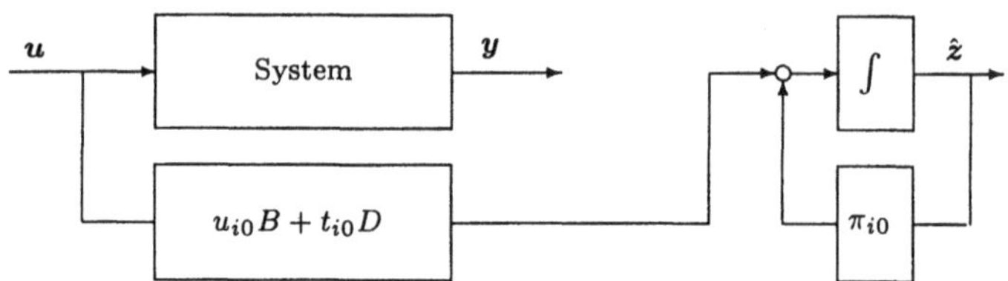

Figure 1.9.   Elementary observer of $z = u_i x$ where u_i is a left eigenvector of A.

projection of u_{i0}), insures good robustness when the observer is used in a feedback loop (see Figure 1.10)[7].

Let us consider the observation of $u_i x$ where u_i is the projection of a left eigenvector u_{0i}. The row vector u_i is in the row span of $U(\lambda_i)$ (see (1.13)) then the elementary observer (λ_i, u_i, t_i) obtained by projection is:

$$\eta_i = u_{i0}(U(\lambda_i)U(\lambda_i)^T)^{-1}U(\lambda_i) \tag{1.69}$$

$$u_i = \eta_i U(\lambda_i) \tag{1.70}$$

$$t_i = \eta_i T(\lambda_i) \tag{1.71}$$

1.2.7.7   Unknown input observers

The decoupling ideas of §1.1.6 are dualized considering an "unknown input observer". When unknown inputs (d below) appear on the measurement and on the state derivative of System (1.1):

$$\dot{x} = Ax + Bu + E'd$$
$$y = Cx + Du + F'd$$

the observation can be made independently of d provided that the design of the observer is such that

$$[u_i \quad t_i] \begin{bmatrix} A - \pi_i I & E' \\ C & F' \end{bmatrix} = 0 \tag{1.72}$$

This result is justified in the appendix, see page 279.

This decoupling approach for observer design will be used in the next section for solving the Exact Loop Transfer Recovery Problem.

1.2.7.8   Summary

Designing an observer consists of building the matrices U, T and Π satisfying $UA + TC = \Pi U$. In view of (1.59), these matrices can be built by selecting some triples (π_i, u_i, t_i) as in Equation (1.58) (there is no limitation on the number of selected triples). Depending on the design objectives, each assignment can be subject to additional constraints which are

- Equation (1.72) for decoupling (unknown input observer).

- Equations (1.70)-(1.71) for projection.

Note that when π_i is non-real, the triple $(\overline{u}_i, \overline{t}_i, \overline{\pi}_i)$ must also be selected so that a simple transformation permits us to manipulate matrices U, T and Π having real entries (see page 49). The function ob_gene performs these assignments, it consider automatically complex conjugate values and transforms the matrices U, T, Π to real matrices.

The vector of observed variable is $z = Ux$, its estimate is $\hat{z}$ (see Figure 1.7). Two matrices Q_y and Q_z (see page 50) permit the designer to observe any linear combinations of the entries of y and $\hat{z}$.

For observers with the special Kalman filter structure, see page 50.

Recommendation. *Alike in the feedback case (page 30), for selecting the triples (π_i, u_i, t_i), it is proposed to proceed by continuity from the open-loop left eigenstructure. But here, there is a major difference: The selected vectors u_i, must define observation as complementary as possible to the measurements. So, the angles between the row span of C and the selected vectors u_i must be as large as possible (the function ob_ins performs such optimization).*

Software

Related MATLAB-**functions**: ob_gene, sob_proj ob_ins and add_obs. The three first functions compute the matrices U, T and Π (see (1.59) and Figure 1.7). Using ob_gene, for each triple (π_i, u_i, t_i) it is possible to consider Equations (1.70)-(1.71) or Equation (1.72). The function sob_proj uses internally projection ideas for a systematic design of a full state observer. It also computes the gain K_o of Figure 1.8. The function ob_ins is used for maximizing the angles between the vectors u_i and the rows of the matrix C. The function add_obs connects an observer obtained *via* ob_gene, sob_proj or ob_ins to the system as depicted in Figure 1.7. The matrix $[Q_y \; Q_z]$ is denoted Q in the software. The values that can be given to Q are discussed on page 50.

Example. For a given system **sys** (two measurements), we would like to have two additional observations. For the first observation, we want an eigenvalue **pi1** equal to $= -1.6$ and the third entry of u_1 equal to zero. For the second one, we want to project the open-loop left eigenvector corresponding to the open-loop eigenvalue -0.1837 an move the corresponding eigenvalue to **pi2** $= -0.8$. Definition of the system **sys** having two outputs:

```
sys = [1 0 0 0;0 0 0 1]*rcamdata('lat',0,1);
```

Observer design:

```
pi1 = -1.6; keyo(1) = 'z'; def_pbo(1) = [3];
pi2 = -0.8; keyo(2) = 'p'; def_pbo(2) = [-0.1837];
[U,T,Pi] = ob_gene(sys,[pi1 pi2],keyo,def_pbo);
```

key(1) = **'z'** stands for zero assignment of some entries of u_2 (and/or t_2). **def_pb(1,1)** defines the entry of u_2 that must be zero.

key(2) = **'p'** stands for open-loop left eigenvector projection. **def_pb(1,2)** gives the open-loop pole reference for the assignment of **pi2**.

Note that an alternative observer design consists of using **ob_ins** instead of **ob_gene**. It will result in an optimization of the angles between the observed variables.

```
[U,T,Pi] = ob_ins(sys,[pi1 pi2]);
```

Finally it is possible to connect the observer and the system as in Figure 1.7 using:

```
sysob = add_obs(sys,U,T,1);
```

The fourth input argument being equal to 1, it means that the matrix $[Q_y \; Q_z]$ of Figure 1.7 is the identity matrix. Therefore, the outputs of **sysob** are y and $\hat{z}$.

1.2.8   Observer-based output feedback

Observer-based output feedback design is usually presented as two independent steps: the observer design step plus a *state feedback* design step. Here, it is suggested to first design the observer and then to consider the observed variables as additional measurements (see (1.74)) for output feedback design. It is not useful to observe signals that are *a priori* given, or to observe the whole state vector. In fact, observers add degrees of freedom for control design. So, it is suggested to add just as many observations as necessary to make it possible to control all the dominant poles (see page 30), ignoring which signals are observed.

1.2.8.1 Implementation of an observer-based control law

The computation of the feedback gains K_y and K_z will be treated later, together with the separation principle. Figure 1.10 shows the structure that is considered here for observer-based output feedback *implementation*. The closed-loop equation of the system of Figure 1.10 is:

$$\dot{x} = Ax + Bu$$
$$\dot{\hat{z}} = \Pi\hat{z} + (UB + TD)u - Ty$$
$$y = Cx + Du$$

with feedback

$$u = K_y y + K_z \hat{z}$$

It can be written in two forms. The first form (1.73) is suitable for implementing an output feedback (*i.e.*, K_y and K_z designed by the technique of §1.2.1). Assuming that K_y and K_z are given, the second form gives the corresponding dynamic feedback $u = G(s)y$ which can be implemented directly on System (1.1).

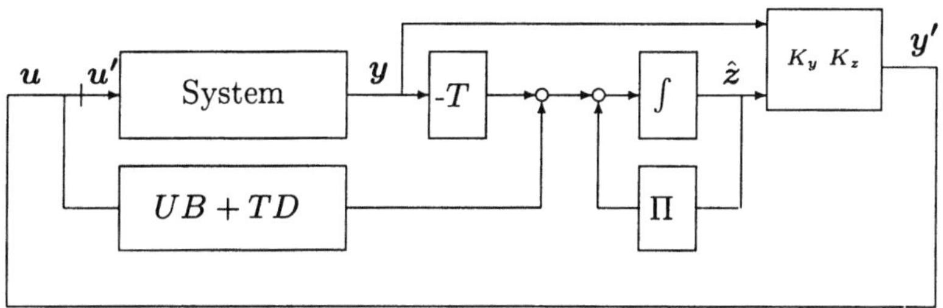

Figure 1.10. Observer-based closed-loop system.

First form: We have

$$\begin{bmatrix} \dot{x} \\ \dot{\hat{z}} \end{bmatrix} = \begin{bmatrix} A & 0 \\ -TC & \Pi \end{bmatrix} \begin{bmatrix} x \\ \hat{z} \end{bmatrix} + \begin{bmatrix} B \\ UB \end{bmatrix} u$$
$$\begin{bmatrix} y \\ \hat{z} \end{bmatrix} = \begin{bmatrix} C & 0 \\ 0 & \Pi \end{bmatrix} \begin{bmatrix} x \\ \hat{z} \end{bmatrix} + \begin{bmatrix} D \\ 0 \end{bmatrix} u \tag{1.73}$$

with feedback

$$u = \begin{bmatrix} K_y & K_z \end{bmatrix} \begin{bmatrix} y \\ \hat{z} \end{bmatrix} \tag{1.74}$$

Second form: Figure 1.10 can be interpreted as shown in Figure 1.11 so that the observer-baser controller can be viewed as standard control

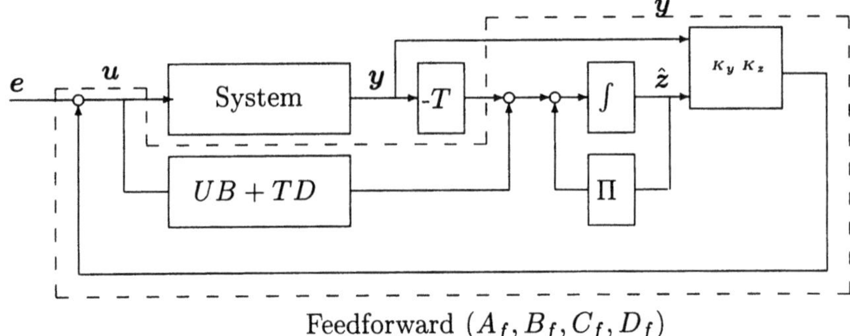

Feedforward (A_f, B_f, C_f, D_f)

Figure 1.11.   Observer-based controller.

system with a feedforward and a feedback gain.   The original system
(A, B, C, D) is controlled by the dynamic gain (transfer function matrix
from $\boldsymbol{y}$ to $\boldsymbol{u}$) $G(s) = C_c(sI - A_c)^{-1}B_c + D_c$ where:

$$
\begin{aligned}
A_c &= \Pi + (UB + TD)K_z \\
B_c &= -T + (UB + TD)K_y \\
C_c &= K_z \\
D_c &= K_y
\end{aligned}
\tag{1.75}
$$

and the feedforward gain is $F(s) = C_f(sI - A_f)^{-1}B_f + D_f$ (transfer
function matrix from $\boldsymbol{e}$ to $\boldsymbol{u}$) where:

$$
\begin{aligned}
A_f &= \Pi + (UB + TD)K_z \\
B_f &= UB + TD \\
C_f &= K_z \\
D_f &= I
\end{aligned}
\tag{1.76}
$$

1.2.8.2   Feedback design and Separation Principle.

The Separation Principle is a well known result that needs to be re-
stated here as we do not use the standard notation of [Luenberger, 1966].
Before an intermediate design problem is defined.

When a bank of elementary observers is considered, the "measurements" at our disposal can be assumed to be

$$
\begin{bmatrix} y \\ z \end{bmatrix} = \begin{bmatrix} C \\ U \end{bmatrix} x + \begin{bmatrix} D \\ 0 \end{bmatrix} u
\tag{1.77}
$$

instead of y and $\hat{z}$ (see Figure 1.10). So, we can go back to §1.2.1 and design an output feedback gain

$$K = [\ K_y \quad K_z\ ]$$

relative to the system

$$\dot{x} = Ax + Bu$$

$$\left[\begin{array}{c} y \\ z \end{array}\right] = \left[\begin{array}{c} C \\ U \end{array}\right] x + \left[\begin{array}{c} D \\ 0 \end{array}\right] u \tag{1.78}$$

The implemented control law will be as in (1.74):

$$u = [\ K_y \quad K_z\ ] \left[\begin{array}{c} y \\ \hat{z} \end{array}\right] = K_y y + K_z \hat{z}$$

This design procedure is justified in the following lemma.

LEMMA 1.2.8 *"Separation Principle". Assume that*

1. *A n_c order observer is available i.e., three matrices $U \in \mathbb{R}^{n_c \times n}, T \in \mathbb{R}^{n_c \times p}$ and $\Pi \in \mathbb{R}^{n_c \times n_c}$ satisfying (1.60) are given. The rows of U are denoted u_i.*

2. *Two feedback matrices $K_y \in \mathbb{R}^{m \times p}$ and $K_z \in \mathbb{R}^{m \times n_c}$ relative to (1.78) are designed such that $p+n_c$ pairs of eigenvalues / eigenvectors λ_i, v_i are assigned (by the way n pairs are assigned).*

Then U, T, Π, K_y and K_z being used as depicted in Figure 1.10, the corresponding closed-loop system is such that:

1. *The eigenvalues of Π belong to the closed-loop spectrum. The corresponding left eigenvectors are $[u_i \quad u_{ic}]$ for some vector $u_{ic}^* \in \mathbb{C}^{n_c}$.*

2. *The n eigenvalues of System (1.78) controlled by K_y and K_z belong to the closed-loop spectrum. In addition the corresponding right eigenvectors are $[v_i^* \quad v_{ic}^*]^*$ for some vector $v_{ic} \in \mathbb{C}^{n_c}$.*

Proof. See Appendix 4, page 282.

1.2.8.3 Summary

There are two steps for designing an observer-based feedback law: observer design and then output feedback (or state feedback) design.

- The observer order should be chosen as being equal to the difference between the number of dominant poles that need to be assigned and the number of measurements. See the summary of page 53.

- Then design a feedback as in §1.2.1 considering indifferently Systems (1.73) or (1.78). See the summary of page 30.

Furthermore, the transformations of Equations (1.75) and (1.76) can be applied to the result if the observer structure is not relevant.

Software

Related MATLAB-**functions**: `add_obs`, `obs2dfb`, `dfb_ins` and `dfb_proj`. The first function realizes (1.73), the second one (1.75). The two other functions are systematic state observer-based design functions, `dfb_ins` optimizes the angles between the eigenvectors (see page 35) and `dfb_proj` uses projection ideas (from open-loop, see §1.1.7).

Example of an observer based feedback design. For the system including the observer of page 54 we would like to design a feedback. First, a feedback gain `K1` relative to the triple (`A,B;[C;U]`) is designed. Second, a feedback gain `K2` relative to System (1.73) is designed. The results must be identical. The observer design of page 54 is first recalled.

```
    sys = [1 0 0 0;0 0 0 1]*rcamdata('lat',0,1);
```
Design of `K1`
```
    C1 = [sys.c;U]; D1 = [sys.d;zeros(2,2)];
    sysob1 = ss(A,B,C1,D1);
    pol_ol = [-0.23+0.59i,-0.18,-1.3];
    pol_cl = [-0.60+0.60i,-0.60,-1.3];
    K1 = fb_prop(sysob1,0,pol_cl,'m',pol_ol)
```
Design of `K2`
```
    sysob2 = add_obs(sys,U,T,1);
    K2 = fb_prop(sysob2,0,pol_cl,'m',pol_ol)
```
It can be checked that `K1` = `K2`. From the separation principle, the spectrum `eig_fb(sysob2,K2)` is equal to { `pi1`, `pi2` } ∪ `eig_fb(sysob1,K1)`. **Other examples:** see pages 136, 137 and the on-line `help2` message of `sob_proj`.

1.2.9   Observer-based state feedback

State feedback is a special case of output feedback when the direct transmission D is equal to zero. So, the previous discussion relative to output feedback holds to some extent for state feedback. The specificity of state feedback are treated here, in particular, the Exact Loop Transfer Recovery problem is considered.

1.2.9.1 Implementation of a state feedback

As shown in (1.65), when D is non-zero, a matrix $Q_u = -Q_y D$ must be considered in order to obtain an estimate of the state vector. The proposed toolbox does not consider observers with non-zero matrix Q_u. For this reason, we derive here a state feedback implementation procedure that is consistent with the observer structure of Figure 1.10.

Let $\widetilde{K}$ denote a given state feedback. We would like to implement it considering System (1.78) (from the Separation Principle, the results will remain valid considering System (1.73)). Therefore, two matrices Q_y and Q_z are computed such that

$$\begin{bmatrix} Q_y & Q_z \end{bmatrix} \begin{bmatrix} C \\ U \end{bmatrix} = I \tag{1.79}$$

then, multiplying on the left the measurement equation by $[Q_y \quad Q_z]$, System (1.78) can be written

$$\begin{aligned} \dot{x} &= Ax + Bu \\ \begin{bmatrix} Q_y & Q_z \end{bmatrix} \begin{bmatrix} y \\ z \end{bmatrix} &= x + Q_y Du \end{aligned} \tag{1.80}$$

The feedback that must be applied to System (1.80) modified as above is not $\widetilde{K}$ but the equivalent feedback that takes the direct transmission $Q_y D$ into account, that is (see (4.3), page 247)

$$K = (I + \widetilde{K} Q_y D)^{-1} \widetilde{K}$$

From the Separation Principle, it is equivalent to apply this feedback either to System (1.78) ($u = K(Q_y y + Q_z z)$) or to System (1.73) ($u = K(Q_y y + Q_z \hat{z})$).

In conclusion, in order to apply a given state feedback gain $\widetilde{K}$ using the observer structure of Figure 1.10:

1 Compute Q_y and Q_z satisfying (1.79).

2 Compute K from $\widetilde{K}$.

3 Compute $K_y = KQ_y$ and $K_z = KQ_z$.

In the Kalman filter structure case, $Q_y = 0$, so, we have $K = \widetilde{K}$ and then, $K_y = 0$, $K_z = \widetilde{K} U^{-1}$.

1.2.9.2   Exact Loop Transfer Recovery.

First, let us briefly recall what means "Loop Transfer Recovery" (LTR). This concept was defined in [Doyle and Stein, 1979, Doyle and Stein, 1981]. It is assumed that a state feedback is available and that this feedback presents some robustness properties related to the shape of the return difference at the plant input. But if this state feedback is implemented using a standard observer, the return difference at the plant input is modified and the related robustness properties are lost. An observer which satisfies the LTR property is such that, if the loop is broken at the plant input, the transfer is the same in both state and observer-based feedback cases. An *asymptotic* solution to this problem is usually proposed ([Doyle and Stein, 1979, Doyle and Stein, 1981]).

Here, a less general *exact* solution is proposed (see [Apkarian et al., 1989] for details). It is assumed that $D = 0$. The proposed technique consists of designing an unknown input observer in which the unknown input is $\boldsymbol{u}$, so, in view of Equation (1.72), it suffices to choose u_i and t_i such that $u_i B = 0$. Therefore, the design of the observer is made by solving this equation together with (1.58), *i.e.*,

$$[u_i \quad t_i] \begin{bmatrix} A - \pi_i I & B \\ C & 0 \end{bmatrix} = 0 \qquad (1.81)$$

The LTR problem has an exact solution when it is possible to find a matrix U (the rows of which are the vectors u_i) that satisfy Equation (1.81), and such that for a given state feedback gain K there exists two matrices K_y and K_z satisfying

$$[\ K_y \quad K_z\ ] \begin{bmatrix} C \\ U \end{bmatrix} = K$$

An illustrative example can be found in §2.3.4.

Justification. If the gains K_y an K_z are applied as in Figure 1.10, from Lemma 1.2.6, the transfer between $\boldsymbol{u}$ and $\boldsymbol{y}'$ is

$$\boldsymbol{y}'(s) = [\ K_y \quad K_z\ ] \begin{bmatrix} C(sI - A)^{-1}B \\ U(sI - A)^{-1}B \end{bmatrix} \boldsymbol{u}(s) = K(sI - A)^{-1}B\ \boldsymbol{u}(s)$$

But we would like this property to be satisfied from $\boldsymbol{u}'$ to $\boldsymbol{y}'$ (the loop is broken at the vertical bar in Figure 1.10). So, the LTR property is satisfied if $\boldsymbol{u}' = \boldsymbol{u}$ *i.e.*, $UB + TD = 0$ or $UB = 0$ because $D = 0$.

In order to increase the number of vector u_i satisfying (1.81), it is necessary to consider the "stable" zeros of the systems (A, B, C) as elementary observer dynamics (*i.e.*, values of π_i) (the knowledge of almost zeros as in §1.5.3 is also useful).

In conclusion, for LTR implementation of a given state feedback K ($D = 0$), proceed as follows:

1 Computes $n - p$ triples (π_i, u_i, t_i) satisfying (1.81). Note that, if $p \geq m$, if all the zeros of (A, B, C) belong to the left half complex plane, and if the selected π_i's contain all the zeros of (A, B, C), then, generically the matrix $[C^T U^T]$ is invertible.

2 Implement the state feedback as explained in the previous paragraph (Q_y and Q_z can be computed as in (1.79) because the matrix $[C^T U^T]$ is generically invertible).

For the properties of genericity evoked above, the reader is referred to [Magni, 1990].

1.3 Eigenstructure assignment II: Multi-model approaches

Contents. The multi-model case with structured dynamic gain is considered here. The following points are addressed:

- Dynamic feedback with given structure .... §1.3.2

- Frequency domain constraints .... §1.3.3

- Multi-model eigenstructure assignment .... §1.3.4

- Multi-model "phase control" .... §1.3.5

- Controller order reduction .... §1.3.6

- Illustrative examples .... §1.3.7 and 1.3.8

References. Most of the theory presented here is extracted from [Magni, 1999, Magni et al., 1998, Magni et al., 1997a]. Realistic applications can be found in [Chiappa et al., 1998, Le Gorrec et al., 1997, Le Gorrec et al., 1998b, Magni et al., 1997a]. A combination of μ-analysis and multi-model design is proposed in [Magni et al., 1998]. An adaptation of the proposed multi-model design technique to direct scheduled control design can be found in [Magni, 1999] and an application is proposed in [Döll et al., 2000].

1.3.1 Introduction

In §1.2.6 was recalled a traditional approach for designing a dynamic output feedback. Such an approach used together with the technique described in §1.2.1, requires the assignment of all the entries of the eigenvectors. This is not useful in practice as explained now.

Let us denote the controller in state space form as: (A_c, B_c, C_c, D_c), its transfer matrix is $G(s) = C_c(sI - A_c)^{-1}B_c + D_c$. With the output of the compensator connected to the input of the system and the output of the system connected to the input of the compensator the global state space representation is ($D = 0$ for simplifying notations):

$$\begin{bmatrix} \dot{x} \\ \dot{x}_c \end{bmatrix} = \begin{bmatrix} A + BD_cC & BC_c \\ B_cC & A_c \end{bmatrix} \begin{bmatrix} x \\ x_c \end{bmatrix} \tag{1.82}$$

let λ_i be an eigenvalue of this matrix and the corresponding right eigenvectors of the closed-loop system are of the form:

$$\begin{bmatrix} v_i \\ v_{ci} \end{bmatrix}$$

Usually, the designer must shape some "regulated outputs". Let us denote z such a vector. It is defined as $z = Ex$. Considering the modal decomposition of (1.19)

$$z = \sum_{i=1}^{i=n} Ev_i\xi_i(t) \tag{1.83}$$

Traditional eigenstructure assignment consists of shaping the vectors Ev_i in order to control the distribution of the modes on the controlled outputs or in order to reproduce open-loop behavior. In a similar way, when a dynamic feedback, *i.e.*, a dynamic extension of the state space, is considered:

$$z = \sum_{i=1}^{i=n+n_c} \begin{bmatrix} E & 0 \end{bmatrix} \begin{bmatrix} v_i \\ v_{ci} \end{bmatrix} \xi_i(t) = \sum_{i=1}^{i=n+n_c} Ev_i\xi_i(t)$$

The vectors v_{ci} disappear in the formula giving the regulated outputs z, so they can be ignored. Taking into account the fact that it is difficult to choose them on the basis of physical considerations, not only they *can* be ignored but they *must* be ignored. This is the main difference between the way we use dynamic extension and the more traditional approaches of §1.2.6.

1.3.1.1 Main result

The following lemma states that we can assign triples (v_i, w_i, λ_i) satisfying (1.6) by solving an equation similar in appearance to (1.7) except that λ_i appears in the equation (K is replaced by $G(\lambda_i)$).

LEMMA 1.3.1 *A triple (v_i, w_i, λ_i) which satisfies (1.6) can be assigned in closed-loop by any dynamic gain $G(s)$ satisfying*

$$G(\lambda_i)(Cv_i + Dw_i) = w_i \tag{1.84}$$

Proof. See Appendix 4, page 250.

Comment. This result suggests to define closed-loop eigenvalues and right eigenvectors as follows. The "*closed-loop eigenvalues*" are the values of λ at which the matrix $A + BG(\lambda)(I - DG(\lambda))^{-1}C$ is rank deficient. The vectors such that $(A + BG(\lambda)(I - DG(\lambda))^{-1}C)v = 0$ are the corresponding "*closed-loop right eigenvectors*" (note that there are $n + n_c$ pairs of such eigenvalues / eigenvectors in the case of a dynamic feedback of order n_c). Lemma 1.3.1 is a result that solves the problem of eigenstructure assignment as revisited in this comment.

1.3.1.2 Linear equality constraints

For solving the problem of Lemma 1.3.1 we will write $G(s)$ in transfer function matrix form:

$$G(s) = \begin{bmatrix} b_{011} + \dfrac{b_{111}s^{q-1}+\ldots+b_{q11}}{s^q+a_{111}s^{q-1}+\ldots+a_{q11}} & \cdots & b_{01p} + \dfrac{b_{11p}s^{q-1}+\ldots+b_{q1p}}{s^q+a_{11p}s^{q-1}+\ldots+a_{q1p}} \\ \cdots & \cdots & \cdots \\ \cdots & \cdots & \cdots \\ b_{0m1} + \dfrac{b_{1m1}s^{q-1}+\ldots+b_{qm1}}{s^q+a_{1m1}s^{q-1}+\ldots+a_{qm1}} & \cdots & b_{0mp} + \dfrac{b_{1mp}s^{q-1}+\ldots+b_{qmp}}{s^q+a_{1mp}s^{q-1}+\ldots+a_{qmp}} \end{bmatrix} \tag{1.85}$$

(Note that the toolbox offers the possibility of choosing distinct degrees for the denominator of the entries of $G(s)$.)

Assuming that the denominators are chosen *a priori*, by substitution of (1.85) into (1.84), *linear equality constraints relative to the numerator coefficients b_{ijk} are obtained*. This point will be developed further from §1.3.2 to §1.3.5.

Comment. It is not very restrictive to chose *a priori* the denominator coefficients. Numerator coefficients have much more effect on performance. This fact is easily interpreted by considering a dynamic compensator as a proportional compensator relative to the natural measurements but also to additional artificial measurements that are the pseudo-derivatives of the natural measurements. In that way, the denominator is required for filtering derivatives (leading to pseudo-derivative with bandwidth depending on noise characteristics). The choice of the denominator coefficients must also take into account the expected feedback bandwidth. In practice, it is often verified that varying the denominator poles of even more than 20 % has almost no effect on performance (after numerator coefficient are recalculated). However, for flexible control, it is better to locate some denominator poles close to the oscillatory poles that must be controlled and for integral effect, it is necessary to consider denominator poles close to the origin.

1.3.1.3   The quadratic criterion

When a dynamic feedback is used as above, the number of degrees of freedom (*i.e.*, number of b_{ijk}) exceeds usually the minimum number that is required for eigenvector assignment. In order to control exceeding degrees of freedom it is proposed to keep the feedback gain as close as possible to a reference one (that might be the null feedback) denoted G_{ref}. So, the following criterion will be considered.

$$J = \sum_{i=1,\ldots,r} \|G_{\text{ref}}(j\omega_i) - G(j\omega_i)\|_F^2 \tag{1.86}$$

where $\omega_i, i = 1, \ldots, r$ are frequencies that are to be chosen in the bandwidth where it is expected that the gain follows the reference gain. It is shown in Appendix 4, page 252, that *this criterion is quadratic with respect to the coefficients* b_{ijk}.

1.3.1.4   Linear Quadratic Programming

Eigenvector assignment will be shown (§1.3.4) to correspond to linear *equality* constraints relative to the numerator coefficients. So, considering a quadratic criterion, the problem to be solved is a *Least Squares problem*.

Dynamic feedback with given structure (see §1.3.2), frequency domain templates (see §1.3.3) and multi-model "phase control" (see §1.3.5) are additional linear equality or *inequality* constraints. The resulting problem becomes, in case of *inequalities*, a *Linear Quadratic Programming problem* that is:

$$J = \frac{1}{2} \xi^T H \xi + f^T \xi \tag{1.87}$$

where ξ satisfies

$$\begin{aligned} Q_1 \xi &= R_1 \\ Q_2 \xi &\leq R_2 \end{aligned} \tag{1.88}$$

in which ξ is the column vector of the parameters b_{ijk}, "$\leq$" is an element-wise inequality, H is positive definite matrix, f^T is a row vector. The matrices H, f, Q_1, Q_2, R_1, R_2 are automatically computed by the tool-box functions. The transformation between the vector ξ and the transfer function matrix are also treated automatically (see §4.1.4, page 251 for details).

Finally, the user may ignore that the software solves a Linear Quadratic Programming problem as only higher level concepts are to be considered for design.

Software

Related MATLAB-**functions**: fb_dyn, eig_cstr, str_cstr, dp_cstr, add_cstr and ktf_crit. The main function is fb_dyn, the specificity of other functions is discussed in the next subsections. Let us just explain the principle of the use of fb_dyn. This function admits two main arguments that are two matrices denoted CRIT and CSTR. These matrices are built by invoking the aforementioned functions. More precisely:

- str_cstr defines the gain structure (choice of the denominators of $G(s)$ and linear constraints relative to the numerator coefficients). This function must be first invoked, in order to initialize the matrix CSTR.

- ktf_crit defines the criterion (so, creates the matrix CRIT).

- Then, the following functions can be invoked one or more times, without specific order. Each time one of these function is invoked, some new constraints are "plugged" into the matrix CSTR.

  - eig_cstr for eigenstructure assignment.

  - dp_cstr for "phase control".

  - add_cstr for frequency domain constraints.

The transfer matrix form of the resulting feedback can be displayed to the screen by using fb_view.

1.3.2 Dynamic feedback with given structure

Each gain entry in (1.85) can be structured as follows:

- The denominator (degree and coefficients) is chosen.

- The degree of the numerator can be chosen, in addition linear equality or inequality constraints can be defined for each numerator coefficient.

For example, a feedback gain may have the following form:

$$G(s) = \begin{bmatrix} 0 & \xi_1 + \frac{\xi_2 s + \xi_3}{s^2 + 2s + 2} \\ \frac{\xi_4 s + \xi_5}{s^3 + 3s^2 + 3s + 1} & \xi_6 \end{bmatrix} \quad ; \quad -1 < \xi_1 < 1$$

in which design parameters are $\xi_1 \ldots \xi_6$. In this example, some b_{ijk} are set to zero, the one corresponding to ξ_1 satisfies two inequality constraints.

More generally, giving a structure to the controller corresponds to defining equality or inequality constraints that are consistent with the Linear Quadratic Problem formulation (1.88).

Software

Related MATLAB-**function**: `str_cstr`. This function adds automatically to the matrix `CSTR` (see page 67) the linear constraints that are discussed in this subsection.

1.3.3   Definition of a frequency domain template

By fixing s in the feedback gain it is possible to define additional classes of constraints. We will give two examples.

Example 1. At the frequency $\omega = 0$ Rd/s, all the entries of the feedback gain $G(s)$ are real numbers. Therefore inequalities relative to the steady-state gain can be defined. Two cases are of interest in practice. First, it is often necessary to have high gains at low frequencies. Second, it might also be desirable to reduce the steady-state gain or even to set it to zero. This requirement is encountered for example for active control of a flexible aircraft: An autopilot is designed as a first loop, then a second loop for active structural mode control can be designed. If the low frequency gain of the second loop is constrained to be low, structural mode control will have negligible effect in the autopilot frequency bandwidth.

Example 2. At any non-zero frequency, the gain entries are complex numbers. These numbers can be set to zero (in order to replace notch filters). An alternative constraint of interest consists of limiting the norm of the gain at given frequencies. But usual norms lead to nonlinear constraints. Therefore it is suggested to replace the natural constraint of the form $\|G_{(i,j)}(j\omega_0)\| < M$ ($G_{(i,j)}(s)$ denotes an entry of the gain

$G(s))$ by

$$-\Re(G_{(i,j)}(j\omega_0)) < -M$$
$$\Re(G_{(i,j)}(j\omega_0)) < M$$
$$-\Im(G_{(i,j)}(j\omega_0)) < -M \qquad (1.89)$$
$$\Im(G_{(i,j)}(j\omega_0)) < M$$

By considering a set of frequencies at which such constraints are taken into account, a frequency domain template is defined. The corresponding linear constraints can be written using the notations of Equation (1.88). This specific form is given in Appendix 4, page 260 as a simple special case of the constraints defined on page 71.

```
|  Software  |
```

Related MATLAB-**function**: add_cstr. This function adds automatically to the matrix CSTR (see page 67) the linear constraints that are discussed in this subsection.

1.3.4 Multi-model eigenstructure assignment

By substitution of (1.85) into (1.84), clearly, eigenstructure assignment reduces to linear *equality* constraints relative to the design parameters b_{ijk}. Such constraints may arise from several models provided that they have the same number of inputs and outputs.

Writing these constraints in the form of Equation (1.88) is not straightforward. This problem is treated in Appendix 4, page 252.

For example, considering two models of a given system (A_1, B_1, C_1, D_1) and (A_2, B_2, C_2, D_2), the simplest example of multi-model eigenstructure assignment problem is:

- Choose a triple (λ_1, v_1, w_1) satisfying (1.6) for the first model:

$$\begin{bmatrix} A_1 - \lambda_1 I & B_1 \end{bmatrix} \begin{bmatrix} v_1 \\ w_1 \end{bmatrix} = 0 \qquad (1.90)$$

- Choose a triple (λ_2, v_2, w_2) satisfying (1.6) for the second model:

$$\begin{bmatrix} A_2 - \lambda_2 I & B_2 \end{bmatrix} \begin{bmatrix} v_2 \\ w_2 \end{bmatrix} = 0 \qquad (1.91)$$

- Find $G(s)$ such that

$$\begin{aligned}
G(\lambda_1)(C_1 v_1 + D_1 w_1) &= w_1 \\
G(\lambda_2)(C_2 v_2 + D_2 w_2) &= w_2
\end{aligned} \tag{1.92}$$

In a single-model setting, it is not tempting to assign two times the same vector (which would result in an infinite gain). In the multi-model case, such a problem can be encountered if the models that are considered are close to each other and if the designer does not realize that he is treating twice the "same pole" in both models (for example the Dutch roll poles of an aircraft). So when a new eigenstructure constraint is added to an existing set of constraints, in some cases it might be necessary to release one of the existing ones. Such problems can be avoided by considering models of the same system that are as "far" as possible from each other. In conclusion:

> *When several models of the same system are used simultaneously for eigenvector assignment, much care must be exercised about the compatibility of the eigenstructure assignment constraints.*

Software

Related MATLAB-**functions**: `eig_cstr`. For each model that is considered, `eig_cstr` must be called to select the eigenvectors that are to be assigned (for example, call twice this function for solving respectively (1.90) and (1.91)). The results are plugged into the matrix CSTR.

1.3.5   Multi-model "phase control"

Here we consider controllers that assign the direction of variations of the open-loop eigenvalues when the loop gain (denoted δ below) is tuned. This is a generalization of the SISO approach to phase control of flexible structures.

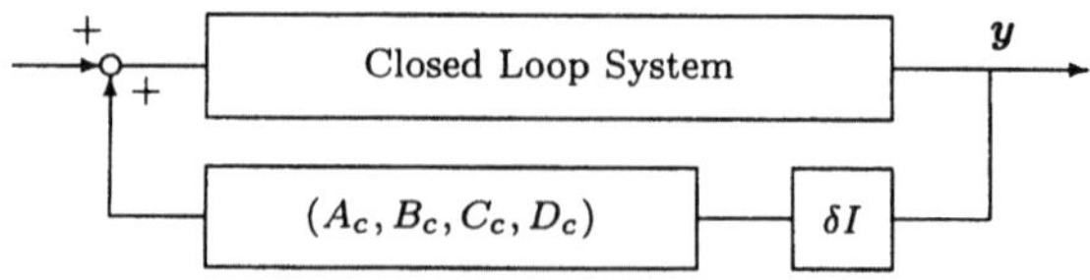

Figure 1.12. Generalized phase control scheme.

Let us denote the controller in state space form as: (A_c, B_c, C_c, D_c)

$$\begin{aligned}
\dot{x}_c &= A_c x_c + B_c y \\
u &= \delta \ (C_c x_c + D_c y)
\end{aligned} \qquad (1.93)$$

δ is the scalar loop gain to be tuned when A_c, B_c, C_c and D_c are known. u is the input vector and y the measurement vector of System (1.1). The key result is:

LEMMA 1.3.2 *For small variations of δ, the variations of the eigenvalues are given (first order approximation) by*

$$\Delta\lambda_i = \delta \ u_i B G_0(\lambda_i) C v_i$$

where $G_0(s) = C_c(sI - A_c)^{-1} B_c + D_c$

Proof. See Appendix 4, page 259.

It is possible to characterize the transfer matrix $G_0(s)$ which are such that $\Re(\Delta(\lambda_i))$ and $\Im(\Delta(\lambda_i))$ satisfy equality and/or inequality constraints. For example:

- $\Re(\Delta(\lambda_i)) \leq \varepsilon$ to move λ_i to the left.

- $\Re(\Delta(\lambda_i)) \leq \varepsilon$ and $\Im(\Delta(\lambda_i)) = 0$, same purpose but pole motion along an horizontal line.

- $\Im(\Delta(\lambda_i)) \leq a\Re(\Delta(\lambda_i))$ and $\Im(\Delta(\lambda_i)) \geq b\Re(\Delta(\lambda_i))$ with $b \geq a$ to restrict pole motion to the left, in a sector with summit at λ_i, defined by two lines of respective slopes a and b.

These constraints can be written as linear constraints of the form of (1.88). See Appendix 4, page 260 for details.

These (first order) constraints can be used together with those defined in §1.3.4. However, some care must be exercised because constraints of §1.3.4 may induce large variations which would invalidate the first order approximations for pole shifting. Note that, in order to avoid this kind of problems, it is possible to use two loops: A low pass loop for eigenstructure assignment (autopilot for example) and a band pass for "phase control" (aircraft structural mode control for example).

Software

Related MATLAB-**function**: `dp_cstr`. This function adds automatically to the matrix CSTR (see page 67) the linear constraints that are discussed in this subsection.

Example: see §1.3.8 (see also [Magni et al., 1997a]).

1.3.6   Controller order reduction

In order to choose the eigenvectors that must be assigned, we mentioned decoupling ideas (see page 21) or projection of open-loop eigenvectors (see page 23). Here, an alternative technique is proposed. It consists of assigning the leading eigenstructure already assigned by a gain which was obtained for example by using $\mathcal{H}_\infty$, μ-synthesis or LQG techniques.

On page 64 is discussed the fact that when a dynamic feedback is used, the state space is extended, but only the part of the eigenvector corresponding to the original system state-space is of interest. The proposed controller order reduction technique consists of re-assigning the relevant part of the dominant eigenvectors. The dominant eigenstructure can be identified using modal simulations (see page 96).

In such a re-design procedure, the reference gain entering in the criterion definition (see (1.86)) is the initial dynamic feedback. The denominators of the structured gain can be chosen taking into account residual analysis of the initial gain (see page 98).

Note that in the same way it is possible to transform a state feedback (LQ design for example) into a dynamic output feedback.

Comment. This very simple technique, used in fact for re-designing a given feedback law in terms of eigenstructure assignment, presents a great advantage as it permits to combine frequency domain techniques ($\mathcal{H}_\infty$, μ-synthesis) with modal approaches.

- On one hand, frequency domain techniques are efficient for frequency domain specifications as robustness with respect to unstructured perturbations, disturbance decoupling, but are not very efficient for dealing with settling times, damping ratio or robustness with respect to *real* parameter variations.

- On the other hand, modal multi-model techniques are efficient for time domain specifications and robustness with respect to real parameter variations, but are not very efficient for frequency domain specifications (however, see §1.3.3).

When, a modal version of a given controller is available, it is possible to

- improve its time domain performance by shifting some poles. Eigenvector are selected by projection of the original closed-loop eigenvectors so that minimum changes are made, preserving in that way the frequency domain performance of the original design.

- to improve robustness by using the multi-model techniques presented in §1.3.4 and §1.3.5. Here again, projections are used in order to induce minimum perturbation of the original performance.

- structure the original feedback gain as in §1.3.2.

Software

Related MATLAB-**function**: **fb_dyn** and related functions. The function for closed-loop pole dominance analysis will be presented later. These functions are: **lsim_mod** (modal simulation, see §1.5.1) is of interest in order to detect visually the dominant poles of the original feedback. Dominant poles can be found more systematically by combining **lsim_mod** (used with output arguments, $\rightarrow \mathcal{L}_2$ dominance), **plot_res** (dominance related to residuals), **plot_con** (dominance related to controllability) and **clean_ev** (for removing duplication after concatenation of several vectors of "dominant" poles).

Example: see page 162 (see also [Chiappa et al., 1998]).

1.3.7 Illustrative example: multi-model assignment

The proposed example considers two models of the same system. A dynamic feedback with entries of the form

$$G_{ij}(s) = * + \frac{*}{s+8}$$

is designed in order to assign six eigenvalues for the first system and four for the second one.

Generation of two models of the same system:

```
sys1 = rcamdata('lat',12,2,0.1);
sys2 = rcamdata('lat',25,2,0.075);
```

Initial feedback for **sys1**:

```
pol0(1) = -0.9+0.9*i; key(1) = 'n'; dfpb(1) = 4;
pol0(2) = -0.8; key(2) = 'n'; dfpb(2) = 4;
pol0(3) = -2.0; key(3) = 'n'; dfpb(3) = 1;
pol0(4) = -1.5; key(4) = 'n'; dfpb(4) = 1;
pol0(5) = -1.0; key(5) = 'n'; dfpb(5) = 1;
K0  = fb_prop(sys1,0,pol0,'n',dfpb);
```

```
fb0 = ss([],[],[],K0);
```

This initial feedback leads to two badly damped closed-loop poles for sys2 which are -2.37 + 4.40i and -1.14 + 3.00i. These poles are shifted to -4 + 4i and -3 + 3i while the initially assigned poles are re-assigned for sys1. Gain structure and criterion:

```
CSTR = str_cstr(2,6,-8,0);
CRIT = ktf_crit(CSTR,fb0,i*[0 0.01 0.05 0.1 0.5]);
```

Eigenstructure assignment for sys1 (projection):

```
oldpol1 = pol0;
newpol1 = pol0;
CSTR = eig_cstr(CSTR,sys1,newpol1,'p',oldpol1,fb0);
```

Eigenstructure assignment for sys2 (projection):

```
oldpol2 = [-2.37+4.40*i -1.14+3.00*i];
newpol2 = [-4+4*i -3+3*i];
CSTR = eig_cstr(CSTR,sys2,newpol2,'p',oldpol2,fb0);
```

Dynamic feedback computation:

```
fb = fb_dyn(CSTR,CRIT);
```

The gain structure can be checked using str_view(CSTR). The assignment can be checked by:

```
eig_fb(sys1,fb)
eig_fb(sys2,fb)
```

1.3.8   Illustrative example: multi-model "phase control"

The previous example illustrates eigenstructure assignment using the functions eig_cstr. Here, using dp_cstr, it is the *pole motion that is assigned* instead of the eigenstructure. Furthermore, the definition of frequency templates using add_cstr is illustrated.

Two systems are generated (1 input, 2 outputs):

```
den1 = [-1+2*i -1-2*i -0.1+3*i -0.1-3*i];
sys1 = ss(zpk({[1];[2]},{den1;den1},[1;2]));
den2 = [-1+4*i -1-4*i -0.1+6*i -0.1-6*i];
sys2 = ss(zpk({[1];[3]},{den2;den2},[1;2]));
```

The gain we look for is strictly proper with a denominator common to all entries. The controllers poles are chosen as $\{-2 \pm 2i, -3 \pm 3i, -5 \pm 5i\}$. For defining the criterion, the reference gain is `fb0` = `[0 0]`:

```
fb0 = ss([],[],[],[0 0]);
```

Gain structure and criterion:

```
CSTR = str_cstr(1,2,[-2+2*i;-3+3*i;-5+5*i],1);
CRIT = ktf_crit(CSTR,fb0,i*[0 0.1 0.5 1 2 5 10]);
```

The constraint defined for example by `pol1(1)` = `-0.1+3*i`; `fld1(1)` = `'r'`; `fld2(1)` = `'<'`; `fld3(1)` = `-0.2`; should be read as follows. The pole at $-0.1 + 3i$ has its real part constrained to reduce at least of -0.2 (when the loop gain δ is equal to 1). The set of constraints defined below for `sys1` is:

- $-0.1 + 3i$ must move to the left at least of -0.2 ($\delta = 1$) together with no variation of the imaginary part.

- $-1.0 + 2i$ must not move.

The set of constraints defined below for `sys2` is:

- $-1.0 + 4i$ must move to the left at least of -0.2 ($\delta = 1$) together with no variation of the imaginary part.

- $-0.1 + 6i$ must not move.

Constraints for pole motion of `sys1`:

```
pol1(1)=-0.1+3*i; fld1(1)='r'; fld2(1)='<'; fld3(1)=-.2;
pol1(2)=-0.1+3*i; fld1(2)='i'; fld2(2)='='; fld3(2)=0;
pol1(3)=-1.0+2*i; fld1(3)='r'; fld2(3)='='; fld3(3)=0;
pol1(4)=-1.0+2*i; fld1(4)='i'; fld2(4)='='; fld3(4)=0;

CSTR = dp_cstr(CSTR,sys1,pol1,fld1,fld2,fld3);
```

Constraints for pole motion of `sys2`:

```
pol2(1)=-1.0+4*i; fld1(1)='r'; fld2(1)='<'; fld3(1)=-.2;
pol2(2)=-1.0+4*i; fld1(2)='i'; fld2(2)='='; fld3(2)=0;
pol2(3)=-0.1+6*i; fld1(3)='r'; fld2(3)='='; fld3(3)=0;
pol2(4)=-0.1+6*i; fld1(4)='i'; fld2(4)='='; fld3(4)=0;

CSTR = dp_cstr(CSTR,sys2,pol2,fld1,fld2,fld3);
```

The frequency domain constraints should be read as follows: `IJ(1,:)` = `[1 1]`; `fl1(1)`= 0; `fl2(1)`= `'<'`; `fl3(1)=0.1`; means that $\|G_{1,1}(0)\| < 0.1$. Both other constraints defined below mean that $\|G_{1,2}(0)\| < 0.1$ and $\|G_{1,1}(100j)\| < 0.1$. (Note that the function **add_cstr** implements the inequalities of (1.89) instead of the above norm constraints.)

```
IJ(1,:) = [1 1]; fl1(1)=   0; fl2(1)= '<'; fl3(1)=0.1;
IJ(2,:) = [1 2]; fl1(2)=   0; fl2(2)= '<'; fl3(2)=0.1;
IJ(3,:) = [1 1]; fl1(3)=100; fl2(3)= '<'; fl3(3)=0.1;

CSTR = add_cstr(CSTR,IJ,fl1,fl2,fl3);
```

Gain computation:

```
fb = fb_dyn(CSTR,CRIT);
```

In conclusion, Figures 1.13 and 1.14 show the expected results. The computation of the gain at 0 and 100 Rd/s would show that the constraints defined by **add_cstr** are saturated.

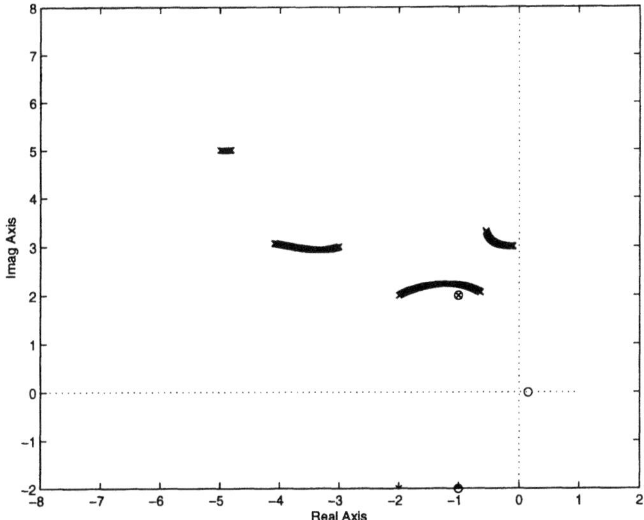

Figure 1.13. Root locus of `sys1`

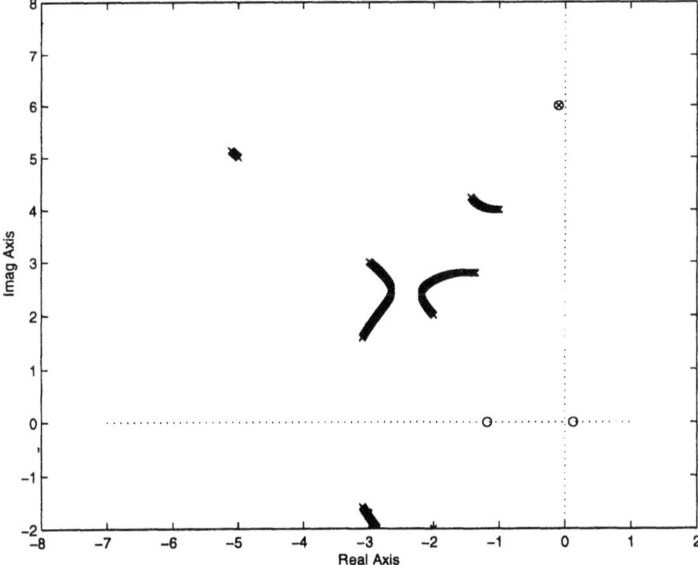

Figure 1.14. Root locus of `sys2`

1.4 Eigenstructure assignment III: Feedback gain tuning

Contents. Tuning is efficient only when a good initial design is available. The technique presented here is a multi-model approach. We will consider:

- Constraints for pole shifting .... §1.4.2

- Constraints for gain structuring .... §1.4.2

- Examples of MATLAB-files to be written by the user .... §1.4.3

- More advanced constraints .... §1.4.4

References. The approach presented here is based on the first order perturbation theory that can be found in [Wilkinson, 1965]. The results from [Wilkinson, 1965] were adapted to a control design purpose in[Magni and Manouan, 1994], this reference introduces also the iterative technique considered in this section. This design technique was assessed in an industrial setting as reported in [Livet, 1995]. Alternative more standard iterative techniques are reviewed in [Liu and Patton, 1998].

1.4.1 Introduction

In order to illustrate briefly the proposed iterative tuning technique, let us consider abstract constraints and a criterion that will be precisely defined later.

Let K_0 be an initial gain. At each step of the proposed tuning procedure, a variation ΔK which satisfies some constraints and minimizes the criterion is computed. Then, K_0 is replaced by $K_0 + \Delta K$, the set of constraints is updated, a new ΔK is computed, and so on.

Step by step, the set of constraints changes in such way that at the final step it corresponds to the eigenstructure assignment problem that must be solved. The updates of the set of constraints must be small enough so that first order approximations remain valid.

1.4.1.1 Linear constraints that will be addressed

The considered constraints are mostly based on first order perturbation theory of eigenstructure (except those of the second set of §1.4.2). The details are presented in §1.4.2 and 1.4.4. Three kinds of constraints will be considered:

- The first set of constraints of §1.4.2 is relative to pole shifting. In order to avoid conflicting objectives, equality constraints must be replaced by inequalities in so far as possible. So, it is proposed to shift some poles of a set of models towards some area of the complex plane, these objectives correspond to *inequality* constraints. Exact pole assignment should be avoided as it corresponds to *equality* constraints. These constraints are relative to ΔK.

- The constraints of the second set of §1.4.2 are global, which means that they are relative to $K_0 + \Delta K$ (instead of ΔK). Exact right eigenvector and interval constraint relative to the entries of the gain are proposed.

- The more general constraints of §1.4.4 are (for some function f) of the form: $f(\lambda_i, v_i, u_i, w_i, t_i)$ is equal to zero or is minimum. The corresponding constraint that will be defined is as follows:

$$f(\lambda_i, v_i, u_i, w_i, t_i) < \tilde{f}$$

in which, for example $f = u_i \Delta A v_i + u_i \Delta B w_i + t_i \Delta C v_i + t_i \Delta D w_i$ (see page 35). The tuning procedure will consist of reducing f step by step.

1.4.1.2 The quadratic criterion

It is recommended to use a criterion relative to the size of the *variation* (ΔK) of the gain at each step (see (crit_k)), that is:

$$J_0 = \|\Delta K\|_F^2 \tag{1.94}$$

The advantage of this criterion is that it induces a small change in the feedback gain, therefore first order approximations appearing in the constraints are likely to be satisfied. If the final objective is to have the gain itself of minimum norm, it is tempting to minimize at each step

$$J_n = \|K_0 + \Delta K\|_F^2 \tag{1.95}$$

but if the initial value of K_0 is not equal to zero, such a criterion might induce large changes which are incompatible with first order approximations. It is better to use a criterion that is close to J_0 at the first iterations and that becomes J_n at the last ones. If i denotes the iteration number and n the total number of iterations, it is suggested to consider

$$J_i = \frac{n-i}{n} J_0 + \frac{i}{n} J_n \tag{1.96}$$

1.4.1.3 Linear Quadratic Programming

Let us define $\xi \in \mathbb{R}^{mp}$. The entries of this vector are the coefficients of the matrix ΔK. Let us denote ΔK_i the ith row of ΔK

$$\Delta K = \begin{bmatrix} \Delta K_1 \\ \vdots \\ \Delta K_m \end{bmatrix} \rightarrow \xi = \begin{bmatrix} \Delta K_1^T \\ \vdots \\ \Delta K_m^T \end{bmatrix}$$

All the control objectives which are considered here, are expressed as *linear equality or inequality constraints* relative to ξ. The proposed criteria (1.94)-(1.95)-(1.96) are *positive quadratic forms* in ξ. Therefore, the problem consist of finding the value of ξ which minimizes the following criterion

$$J = \frac{1}{2}\,\xi^T H \xi + f^T \xi \tag{1.97}$$

where ξ satisfies

$$\begin{array}{rcl} Q_1 \xi &=& R_1 \\ Q_2 \xi &\leq& R_2 \end{array} \tag{1.98}$$

in which "$\leq$" is an element-wise inequality, H is positive definite matrix, f^T is a row vector, Q_1, Q_2 are matrices with mp columns and R_1, R_2 are column vectors.

For the constraints of (§1.4.2) the user may ignore that the software solves a Linear Quadratic Programming problem: the matrices H, f, Q_1, Q_2, R_1, R_2 are automatically computed by the toolbox functions. The transformation between the vector ξ and the gain variation ΔK are also treated automatically.

For the more advanced objectives that can be modeled as equalities or inequalities relative to functions of the form $f(\lambda_i, v_i, u_i, w_i, t_i)$ the matrices H, f, Q_1, Q_2, R_1, R_2 must be defined by the user (see §1.4.4).

Software

Related MATLAB-**functions**: `fb_tun`, `crit_k`, `crit_ctr`, `cstr_dp`, `cstr_ev`, `cstr_qud`, `cstr_eig`, `cstr_k` and `sort_ev`. The main function is `fb_tun`, the specificity of other functions is discussed in the next subsections. Let us just explain the principle of the use of `fb_tun`. This function is invoked sequentially by an algorithm that must be written by the user. At each iteration `fb_tun` admits two main arguments that are the matrices denoted `CRIT` and `CSTR`. These matrices are built by invoking the aforementioned functions. Briefly (see §1.4.3 for details), at each iteration, before calling `fb_tun` the following functions must be combined.

- `crit_k` defines the criterion (so create the matrix `CRIT`). In order to define the criterion J_0 of (1.94) use
  `CRIT = crit_k(ones(size(K0)),zeros(size(K0)));`
  in order to define the criterion J_n of (1.95) use
  `CRIT = crit_k(ones(size(K0)),K0);`.
  In order to define a linear combination of J_0 and J_n define `CRIT` as being the same linear combination of both above "`CRIT`" matrices.

- Initialize the matrix `CSTR` at each step by using `cstr_ini`. Then, the other functions `cstr_..` can be called one or more times without specific order. Each time one of these functions is called, some new constraints are "plugged" into the matrix `CSTR`.

- `sort_ev` is a function that is used to select the eigenvalues that are to be treated each time a function `cstr_..` is called.

1.4.2 Constraints for pole shifting and gain structuring

The constraints which are based on first order perturbation theory are first presented. Global constraints possibly corresponding to large variations of the gain will be considered in a second step. All the constraints

given in this section must be written in the form of Equation (1.98).
The technique used in the toolbox for that purpose is always the same,
it consists of using some properties of the trace operator.

1.4.2.1 Local constraints

Consider the closed-loop system $(\widehat{A}{=}A + B((I - KD)^{-1}K)C)$,

$$\dot{x} = \widehat{A}x$$

Let λ_i be an eigenvalue of the matrix $\widehat{A}$ and v_i, u_i, be the corresponding
right and left eigenvectors.

LEMMA 1.4.1 *Let ΔK denote the variation of the feedback gain K, assuming that u_i and v_i are normalized such that $u_i v_i = 1$*

$$\Delta \lambda_i = (u_i B + t_i D)\Delta K(Cv_i + Dw_i) \qquad (1.99)$$

is the first order variation of the ith closed-loop eigenvalue.

Proof. See Appendix 4, page 255, this is a special case of Lemma 1.2.2
(page 34).

In order to illustrate the computation that is performed in the toolbox,
let us consider a simple example, the generalization to other pole shifting
problems is straightforward. Assume that it is expected to shift the
eigenvalue λ_i to the left of an amount equal to $-\varepsilon$ ($\varepsilon > 0$). We have to
identify the constraint relative to ΔK that corresponds to the inequality

$$\Re((u_i B + t_i D)\Delta K(Cv_i + Dw_i)) < -\varepsilon$$

As this inequality is scalar, it can be written

$$\text{trace}\Re((u_i B + t_i D)\Delta K(Cv_i + Dw_i)) < -\varepsilon$$

or

$$\text{trace}\Re((Cv_i + Dw_i)(u_i B + t_i D)\Delta K) < -\varepsilon$$

or

$$\begin{cases} M = \Re((Cv_i + Dw_i)(u_i B + t_i D)) \\ \text{trace}(M\Delta K) < -\varepsilon \end{cases}$$

This last inequality can be written ($M_{(k,l)}$ and $\Delta K_{(l,k)}$ denotes the entries
of M and ΔK)

$$\sum_{\substack{k = 1,\ldots,p \\ l = 1,\ldots,m}} M_{(k,l)} \, \Delta K_{(l,k)} < -\varepsilon \qquad (1.100)$$

that is in the form of (1.98).

The above computation is performed by the functions `cstr_dp`, `cstr_qud` and `cstr_ev`. The addressed problems are:

- tuning of the real and/or imaginary parts, with possibly pole motion on an "iso-damping" line or inside a sector, see `cstr_dp`,

- shifting the pole towards a quadrilateral, see `cstr_qud`, (illustration See Figure 1.15)

- preserving assigned poles while other properties are tuned, see `cstr_ev`.

1.4.2.2   Local constraints and dynamic extension

The main objective on the following discussion is to explain why it is not relevant to try to improve a given gain by tuning it *after adding degrees of freedom by means of a dynamic extension.*

Assume that a dynamic extension is used (see §1.2.6, page 45). It is assumed that $D = 0$ for simplifying notation, but the proposed discussion remains valid if $D \neq 0$.

Tuning with a dynamic extension means that we look for gain variations of the form

$$\Delta K = \begin{bmatrix} \Delta K_{11} & \Delta K_{12} \\ \Delta K_{21} & \Delta K_{22} \end{bmatrix}$$

The initial left and right eigenvectors are of the form

$$\begin{bmatrix} v_i \\ 0 \end{bmatrix} \text{ and } \begin{bmatrix} u_i & 0 \end{bmatrix}$$

The pole shifting constraint corresponding to (1.99) is

$$\Delta\lambda_i = \begin{bmatrix} u_i & 0 \end{bmatrix} \begin{bmatrix} B & 0 \\ 0 & I \end{bmatrix} \begin{bmatrix} \Delta K_{11} & \Delta K_{12} \\ \Delta K_{21} & \Delta K_{22} \end{bmatrix} \begin{bmatrix} C & 0 \\ 0 & I \end{bmatrix} \begin{bmatrix} v_i \\ 0 \end{bmatrix}$$

so

$$\Delta\lambda_i = u_i B \Delta K_{11} C v_i$$

This first order constraints does not depend on ΔK_{12}, ΔK_{21} and ΔK_{22}. Therefore, these sub-matrices will remain equal to zero if the tuning procedure is initialized by a feedback gain of the form

$$K_0 = \begin{bmatrix} K_{110} & 0 \\ 0 & 0 \end{bmatrix}$$

because this feedback does not change the form of initial right and left eigenvectors. In this case, the tuning procedure cannot take advantage of K_{12}, K_{21} and K_{22}.

This problem disappears if the initial gain is of the form

$$K_0 = \left[\begin{array}{cc} K_{110} & K_{120} \\ K_{210} & K_{220} \end{array} \right]$$

with non-zero sub-matrices K_{120}, K_{210} and K_{220}, because in this case, the initial eigenvectors are not as above and tuning of a dynamic gain becomes relevant. As a conclusion:

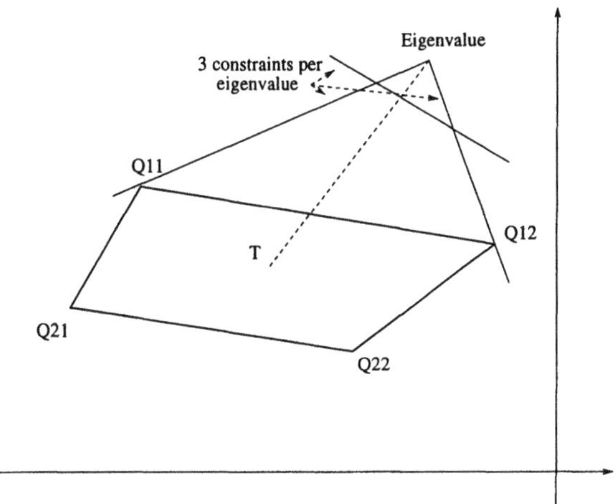

Figure 1.15. Constraints to move one eigenvalue inside a quadrilateral.

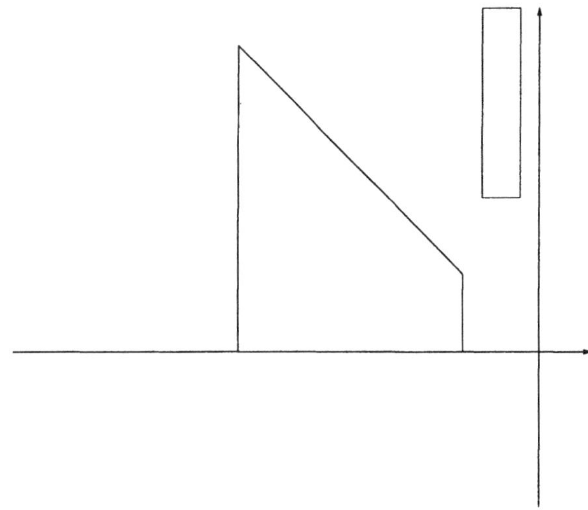

Figure 1.16. Example of relevant quadrilaterals.

If the tuning procedure is considered with an additional dynamic extension, it must be initialized by a gain matrix that does not contain zero-blocks.

Note that tuning an observer-based dynamic gain avoids this kind of problems.

1.4.2.3   Global constraints

At each step of the algorithm, the previous constraints where relative to the gain variation ΔK. Here, the constraints are written in terms of ΔK, but the properties we look for are global and then, hold for (K_0 and) $K_0 + \Delta K$. In general, it must be assumed that the original gain K_0 satisfies the desired global properties. The constraints relative to ΔK must be such that all the updates $K_0 + \Delta K$ made along the tuning procedure preserve the desired global properties. Two cases are considered.

- gain structure: the gains coefficients are the degrees of freedom that are handled. See `cstr_k`.

- eigenvalue / eigenvector assignment: Equation (1.10) is a linear constraint. See `cstr_eig`.

The translation of the constraints relative to the gain into the form of (1.98) is trivial. In order to illustrate the computation that is performed in the toolbox, the problem of eigenstructure assignment is detailed now. Assume that it is expected to find ΔK such that the triple (λ_i, v_i, w_i) is assigned by $K_0 + \Delta K$. The corresponding constraint is

$$(K_0 + \Delta K)(Cv_i + Dw_i) = w_i \tag{1.101}$$

In order to use the trace operator as in the local constraint case, this equation is transformed into m equations. Let e_j denote the m-dimensional row vector with all entries equal to zero except the jth that is equal to one:

$$e_j = [0 \dots 0\ 1\ 0 \dots 0]$$

So, (1.101) is equivalent to a set of m equations of the form

$$e_j(K_0 + \Delta K)(Cv_i + Dw_i)) = e_j w_i, j = 1, \dots, m$$

As these equations are scalar, each one can be written

$$\text{trace}(e_j(K_0 + \Delta K)(Cv_i + Dw_i)) = e_j w_i$$

or

$$\text{trace}(e_j \Delta K (Cv_i + Dw_i)) = e_j(w_i - K_0(Cv_i + Dw_i))$$

or

$$\begin{cases} M_j = (Cv_i + Dw_i)e_j \\ R_j = e_j(w_i - K_0(Cv_i + Dw_i)) \\ \text{trace}(M_j \Delta K) = R_j \end{cases}$$

This last equality can be written

$$\sum_{\substack{k = 1, \ldots, p \\ l = 1, \ldots, m}} M_{j(k,l)} \, \Delta K_{(l,k)} = R_j \tag{1.102}$$

that is in the form of (1.98). The adaptation to complex eigenvector assignment (simultaneous assignment of v_i and $\overline{v}_i$ is straightforward (see page 30)).

Software

Related MATLAB-**functions**: For local constraints (*i.e.*, relative to ΔK), see `cstr_dp`, `cstr_ev`, `cstr_qud`, `cstr_eig`. In addition `cstr_k` which concerns mostly global constraints can also be used for local ones. `dist_qud` is complementary to `cstr_qud`.

The function `cstr_qud` is the most sophisticated as it permits the designer to shift some poles into a given quadrilateral, see Figure 1.15. This figure shows that for each eigenvalue to be shifted towards the quadrilateral (if it is outside), three constraints relative to the gain variation are (internally) defined. Two of them define an allowed sector, the third one defines a minimum shift towards the point T inside the quadrilateral. If an eigenvalue is inside, no constraints are generated. Figure 1.16 illustrates two kinds of quadrilaterals that are relevant for a control design purpose. One, two or more quadrilaterals can be used simultaneously. (Also, as for all other functions of the tuning sub-toolbox, several systems can be dealt with simultaneously.)

The function `cstr_ev` is useful to preserve assigned eigenvalues when, for example, vectors are tuned. The theory behind this function is briefly explained in §3.2.13.

For global constraints (*i.e.*, relative to $K_0 + \Delta K$), see `cstr_k` and `cstr_eig`.

Example: An illustrative example is given in §1.4.3.

1.4.3   Examples of files to be written by the user

The algorithm framework presented here consists of tuning a proportional gains. Adaptations for tuning a dynamic gain are also given. Each m-file to be written by the user for tuning a control law must be structured as follows.

Step 1. Initialization: definition of the system, computation of an initial control law. If the initial feedback is dynamic, transform it into its proportional form

- If a dynamic extension is considered, the gain must be written as in Equation (1.55) (using **dyn2sta**), in parallel, the initial system must be written as in Equation (1.53) (using **add_dyn**) (see page 45).

- If an observer is considered, the initial feedback must be transformed into an equivalent observer-based feedback (using **dfb2obs**), then the observer must be connected to the original system (using **add_obs**, treat preferably the system given as the second output argument of this function).

Let us denote K_0 the initial proportional gain.

Step 2. If the criterion depends only on ΔK, define it here (before the tuning loop) using **crit_k** ($\rightarrow$ CRIT).

Step 3. The tuning loop starts from here. Initialize the constraint matrix using **cstr_ini** ($\rightarrow$ CSTR). If the criterion depends on $K_0 + \Delta K$, define it here using **crit_k** ($\rightarrow$ CRIT).

Step 4. Identify (using **sort_ev**) the eigenvalues assigned by K_0 that must be treated.

Step 5. Treat the eigenvalues identified at step **4** by invoking one of the following functions **cstr_dp**, **cstr_ev**, **cstr_qud**, **cstr_k** or **cstr_eig**. These functions update the matrix CSTR. If additional constraints are to be defined (*e.g.*, multi-model case), go to step **4**, otherwise to to step **6**.

Step 6. Using **fb_tun** compute ΔK that solves the problem defined by CRIT and CSTR. Set $K_0 = K_0 + \Delta K$.

Step 7. Analyze the results using **eig_fb**, **sort_ev**, **dist_qud**. If the final objective is not reached go to step **3**, otherwise, to step **8**.

Step 8. End of the algorithm. If the initial feedback was dynamic, transform K_0 into its dynamic form (using **sta2dyn** in the dynamic extension case or **obs2dfb** in the observer case).

Software

The following example aims at shifting to the left the eigenvalues of a
system so that the real parts become less than -0.5. All eigenvalues with
real part larger than -0.5 are detected using **sort_ev** and are shifted to
the left using **cstr_dp** such that $\Re\Delta\lambda_i < -0.1$.

Step 1.

```
sys = rcamdata('lat',0,1);
kp = 4; km = 2;,
KO = zeros(km,kp);
lambda  = eig_fb(sys);
lambda1 = sort_ev(lambda,'real','>=',-0.5);
```

Step 2.

```
CRIT = crit_k(ones(km,kp),zeros(km,kp));
```

Step 3.

```
while length(lambda1) > 0;

  CSTR = cstr_ini(km,kp);
```

Step 4.

```
lambda2 = sort_ev(lambda,'real','>=',-0.5);
```

Step 5.

```
CSTR =cstr_dp(CSTR,sys,lambda2,'r','<',-0.1,KO);
```

Step 6.

```
dk = fb_tun(CSTR,CRIT);
KO = KO+dk;
```

Step 7.

```
lambda  = eig_fb(sys,KO);
lambda1 = sort_ev(lambda,'real','>=',-0.5);

end;
```

1.4.4   Constraints for more advanced objectives

There is a too wide range of properties that can be written in terms of the vectors v_i, u_i, w_i, t_i. For this reason, it was not possible to develop tools ready for use like in the pole tuning case. This section gives some hints for modifying properties that are written as follows

$$f(\lambda_i, v_i, u_i, w_i, t_i) < \tilde{f}$$

It is proposed to deal with such constraints by using first order perturbation theory (see Lemma 1.4.2 below). More precisely, Δf is written as a function of ΔK. If the initial value of f, say $f_0 = f(\lambda_{i0}, v_{i0}, u_{i0}, w_{i0}, t_{i0})$ is much larger than the objective $\tilde{f}$, in order to avoid abrupt change in the feedback gain (which would be incompatible with first order approximations) it is suggested to proceed as follows. Assuming that $f_0 > \tilde{f} > 0$, at the first step, consider the constraints corresponding to $\Delta f \leq -\varepsilon(f_0 - \tilde{f})$ (with $\varepsilon << 1$, so that f reduces of an amount approximatively equal to $\varepsilon(f_0 - \tilde{f})$, compute a gain variation ΔK in accordance with this inequality, update K_0. Iterating about $1/\varepsilon$ times this procedure, K_0 will become such that the value of f is close to $\tilde{f}$.

Before deriving the expression of the variations Δv_i, Δw_i, Δu_i and Δt_i in terms of gain variation, some notations are defined:

$$V_i = \begin{bmatrix} v_1 & \cdots & v_{i-1} & v_{i+1} & \cdots & v_n \end{bmatrix}$$

$$U_i^T = \begin{bmatrix} u_1^T & \cdots & u_{i-1}^T & u_{i+1}^T & \cdots & u_n^T \end{bmatrix}$$

$$\Lambda_i = \text{Diag}\{\lambda_1, \ldots, \lambda_{i-1}, \lambda_{i+1}, \ldots, \lambda_n\}$$

The following normalizations are taken into account in the lemma formulation: $\|v_i\| = 1$ and $u_i v_i = 1$. It is also assumed that $D = 0$.

LEMMA 1.4.2 *(D = 0) Variations of right eigenvectors.*

$$\begin{aligned} \Delta v_i &= X_v \Delta K Y_v \quad \text{where} \\ X_v &= (I - v_i v_i^*) V_i (\lambda_i I - \Lambda_i)^{-1} U_i B \\ Y_v &= C v_i \end{aligned} \tag{1.103}$$

Variations of input directions.

$$\begin{aligned} \Delta w_i &= X_w \Delta K Y_v \quad \text{where} \\ X_w &= I + KC X_v \end{aligned} \tag{1.104}$$

Variations of left eigenvectors.

$$\begin{aligned} \Delta u_i &= X_u \Delta K Y_u - u_i X_v \Delta K Y_v u_i \quad \text{where} \\ X_u &= u_i B \\ Y_u &= C V_i (\lambda_i I - \Lambda_i)^{-1} U_i \end{aligned} \tag{1.105}$$

Variations of output directions.

$$\begin{aligned}
\Delta t_i &= X_u \Delta K Y_t - u_i X_v \Delta K Y_v t_i \quad where \\
Y_t &= I + Y_u B K
\end{aligned} \qquad (1.106)$$

Proof. See Appendix 4, page 256. Note that when $D \neq 0$, it suffices to replace K by $K(I - DK)^{-1}$ in (1.103) to (1.106).

The tuning procedure defined in §1.4.3 has to be adapted as follows. Steps **4** - **5** become more complex as the user must write himself the constraint relative to the variation of f. It is proposed to write this constraint using the trace operator:

$$\mathrm{trace}(M \Delta K) = \sum_{\substack{k = 1, \ldots, p \\ l = 1, \ldots, m}} M_{(k,l)} \, \Delta K_{(l,k)}$$

The derivation of constraints in this form is already illustrated by the derivation of Equations (1.100) and (1.102). An alternative example is proposed now.

1.4.4.1 Derivation of the matrix M for insensitivity optimization

The criterion of Equation (1.45) (page 35) is relative to the sensitivity of the closed-loop poles. Assuming that the vectors v_i are normalized as in the above lemma, *i.e.*, $\|v_i\| = 1$, this criterion reduces to

$$J = \sum_{i=1}^{n} \|u_i\| = \sum_{i=1}^{n} u_i u_i^*$$

Therefore, the function "$f(\lambda_i, v_i, w_i, u_i, t_i)$" to be treated will be:

$$f = \sum_{i=1,\ldots,n} u_i u_i^*$$

The variation Δf is

$$\Delta f = \sum_{i=1,\ldots,n} u_i \, \Delta u_i^* + \Delta u_i \, u_i^*$$

So, using (1.105) we have

$$\Delta f = 2\Re \sum_{i=1,\ldots,n} X_u \Delta K Y_u - u_i X_v \Delta K Y_v u_i \, u_i^*$$

Δf being scalar, the property $\mathrm{trace}(\Delta f) = \Delta f$ is used for re-ordering

$$\Delta f = 2\Re \sum_{i=1,\ldots,n} \mathrm{trace}\left(X_u \Delta K Y_u - u_i X_v \Delta K Y_v u_i \, u_i^*\right)$$

after re-ordering

$$\Delta f = 2\Re \sum_{i=1,\ldots,n} \mathrm{trace}\left(Y_u X_u \Delta - Y_v u_i u_i^* u_i X_v\right)\Delta K)$$

we will denote

$$M = 2\Re \sum_{i=1,\ldots,n} Y_u X_u \Delta - Y_v u_i u_i^* u_i X_v$$

so

$$\Delta f = \mathrm{trace}\left(M \Delta K\right)$$

It is now possible to define constraints relative to Δf in terms of ΔK. The MATLAB implementation of the whole algorithm is treated below.

Software

When the design objective involves the vectors v_i, u_i, w_i or t_i, step **4** and **5** can be considered as follows:

- Find a matrix M in terms of $X_v, \ldots Y_t$ such that the constraint can be written $\mathrm{trace}(M\Delta K) < -\varepsilon$. Note that `comp_dv` computes the matrices $X_v, \ldots Y_t$.

- Using `k2ksi` transform the above constraint as in (1.98) *i.e.*, $Q_2\xi < R_2$. Note that $\mathrm{trace}(M\ \Delta K)$ is equal to `k2ksi(M')'*k2ksi(dK)` (ξ = `k2ksi(dK)`).

- Add this constraint into the matrix CSTR using `ab2cstr`. If this matrix was not already initialized, use before `cstr_ini`. For some given bound R_2, the constraint `k2ksi(M')'*k2ksi(dK) < R_2` is added to CSTR as follows: `CSTR = ab2cstr(CSTR,[],[],k2ksi(M')',R_2)`

The following example proposes to reduce sequentially the value of f as much as possible with the objective of keeping the poles inside a given trapezium (use of `cstr_qud`). Note that here Steps **4** and **5** are used twice: first time in order to reduce f, second time, in order to keep the poles inside the trapezium. The function `fb_tun` is used with a second output argument for breaking the loop if the problem of reducing f has no longer a solution (minimum reached).

Step 1. (System definition, initial gain, definition of the trapezium (see cstr_qud) and various initializations.)

```
    sys = demodata(1);
    [kp,km,kn] = size(sys);

    pol = [-0.6+0.6i;-1;-0.6];
    KO = fb_prop(sys,0,pol,'z',[4 1 1]);

    Quad = [-10+j*10 -0.25+j*0.25; -10 -0.25];
    Speed = 0.005;

    flg = 'ok';
    jj = 0;
    inv_f = 0;
    Kopt = KO;
```

Step 2. (Criterion definition.)

```
    CRIT = crit_k(ones(km,kp),zeros(km,kp));
```

Step 3. (Constraints re-initialized at each step.)

```
   while flg == 'ok';

     CSTR = cstr_ini(km,kp);
```

Step 4. (Here, all eigenvalues are treated.)

```
    ll = eig_fb(sys,KO);
    [ll,ll2,nll] = clean_ev(ll);
```

Step 5.1. (Value of f tuned towards 0 by reducing its value of f/50 at each step, storage of the best solution in Kopt.)

```
    f = 0;
    M = zeros(kp,km);
    for ii = 1:nll;
       [lam,v,w,u,t,Xv,Yv,Xw,Xu,Yu,Yt]=comp_dv(ll(ii),sys,KO);
       f = f + real(u*u');
       M = M + real(Yu*u'*Xu - Yv*u*u'*u*Xv);
    end;

    if 1/f > inv_f; Kopt = KO; inv_f = 1/f; end;
```

```
CSTR = ab2cstr(CSTR,[],[],k2ksi(M')',-f/50);
```

Step 5.2. (Poles constrained to remain inside the trapezium.)

```
CSTR = cstr_qud(CSTR,sys,Quad,ll,Speed,KO);
```

Step 6.

```
[k,flg] = fb_tun(CSTR,CRIT);
KO = KO + k;
```

Step 7.

```
    if jj == 60; KO = Kopt; break; end;
end;
```

1.5 Modal analysis of a control law

Contents.

- Modal simulation and residuals .... §1.5.1
- Modal controllability .... §1.5.2
- Zeros and "almost zeros" .... §1.5.3

References. An overview of controllability criteria and degree of controllability measures can be found in [Junkins and Kim, 1993] (Chapter 6). The rank test reported in [Junkins and Kim, 1993] was considered for defining a measure of the degree of controllability in [Hamdan and Nayfeh, 1989]. An alternative measure based on energy can be found in [Moore, 1981], but, as this measure cannot be considered for unstable systems, it will be ignored. For modal analysis, we would like to have a correlation between the controllability measure and the propensity of poles to be moved by output feedback. For this reason an "input / output controllability" measure (based on [Hamdan and Nayfeh, 1989]) is proposed in §1.5.2.

An overview of the definitions of the zeros of MIMO systems can be found in [MacFarlane and Karcanias, 1976]. What we call "zeros" here, corresponds to the "invariant zeros" of this reference. Unfortunately, this definition is purely algebraic and is very sensitive to data describing systems. In [Laub and Moore, 1978] a numerical technique is proposed. Here, it is suggested to minimize a singular value in order to find the "zeros" or "almost zeros" depending on the size of the minima. The search of the zeros is reliable as it is initialized by the closed-loop poles corresponding to random high feedback gains.

1.5.1   Modal simulation and residuals

1.5.1.1   Modal simulation

Let us consider the decomposition of the signal obtained in Equation (1.19):

$$\boldsymbol{\xi}_i(t) = e^{\lambda_i t} * u_i B H \boldsymbol{z}_R \quad \text{and}$$
$$\boldsymbol{z} = \sum_{i=1}^{n} [\, E \quad F \,] \begin{bmatrix} v_i \\ w_i \end{bmatrix} \boldsymbol{\xi}_i(t) + \quad \text{direct transmission} \tag{1.107}$$

We call "modal simulation", a simulation in which the components of this sum are simulated separately. Let us denote z_k the kth entry of $\boldsymbol{z}$ and E_k, F_k the kth row of the matrices E and F. It is proposed to simulate (see lsim_mod), the following signals

$$[\, E_k \quad F_k \,] \begin{bmatrix} v_i \\ w_i \end{bmatrix} \boldsymbol{\xi}_i(t) \tag{1.108}$$

In fact, if λ_i is non-real, it is the sum corresponding to λ_i and $\overline{\lambda}_i$ that is considered.

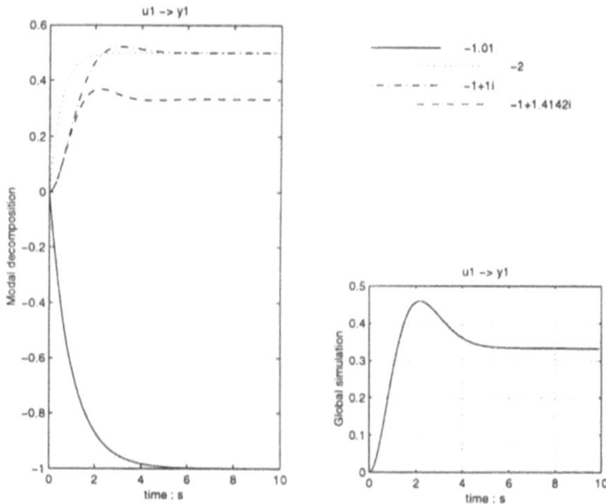

Figure 1.17.   Example of a "modal simulation" (left), corresponding global simulation (right)

Such a tool permits the designer to know exactly which modes are involved in a time response for example when coupling or overshoots are analyzed. Figure 1.17 proposes an illustrative example. The four signals on the left hand side plot are those of Equation (1.108). The sum of the four components is the signal on the right.

However, it is not recommended to define automatically dominant modes because, in some cases, two or more modes that seem to be dominant compensate each other. For example, consider the following academic example in which the first and second transfer functions are almost equal but of opposite sign.

$$S(s) = \frac{1}{s+1} - \frac{1.01}{s+1.01} + \frac{0.1}{s^2+4s+2}$$

The corresponding modal simulation is given in Figure 1.18. In this figures, at first sight, the modes corresponding to the first and second transfer functions seem to be dominant. Clearly, the dominant mode

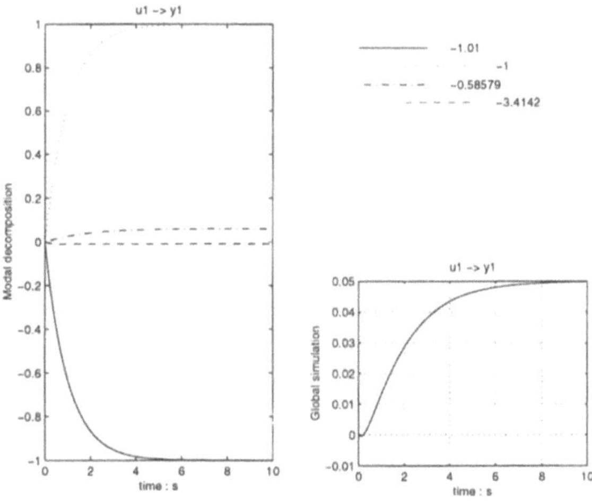

Figure 1.18.   Example of a "modal simulation" (left), corresponding global simulation (right)

correspond to the poles of the third transfer function (that can be easily checked using one of the optional arguments of the function `lsim_mod`). In practice, system states are chosen in such a way that this kind of problems does not arise. However, this problem might be induced by artificial components, for example time delay models.

In conclusion, dominant mode analysis is usually straightforward. In some cases, trials and errors are necessary to remove from appearing dominant modes those that cancel each other.

1.5.1.2   Residuals

When it is necessary to detect the dominant poles in a systematic way,
simulations are time consuming. The modal decomposition of a signal
can also be roughly quantified by computing time response at given instants for a given input. For example, the initial value of the impulse
response or the steady state value of the step response are good candi-
dates. (Note that there is a risk of considering as dominant some modes
that cancel each other, see the comments relative to modal simulations.)

Let us consider the kth output and the lth input. So, in Equation
(1.107), $BH z_R$ is replaced by $B_l \delta$ (impulse response case) or by B_l (unit
step response case).

Case 1: impulse response at time $t = 0$.

$$z_k(t = 0) = E_k v_1 u_1 B_l + \ldots + E_k v_n u_n B_l$$

The quantities $E_k v_i u_i B_l$, $i = 1, \ldots, n$ are called the *residuals* between
input number l and output number k.

Case 2: unit step response at $t \to \infty$.

$$z_k(t \to \infty) = -\frac{E_k v_1 u_1 B_l}{\lambda_1} \ldots - \frac{E_k v_n u_n B_l}{\lambda_n}$$

This second measure of the decomposition of a signal is often more reliable than the first one but systems with eigenvalues at the origin cannot
be considered. The unit step response can also be considered at intermediate instants $(t > 0)$.

Software

Concerned MATLAB-**function**: lsim_mod and plot_res. Figure 1.17
was generated as follows:

```
sys = rcamdata('lat',0,1);
K=fb_prop(sys,0,[-.6+.6*i -.6 -1.3],'n',[3 2 2]);
sysfb=feedback(sys,K,1);

lsim_mod(sysfb,'u2',20,0,'y1',4);
```

This analysis is relative to the first measurement ('y1') for a step input at the second input ('u2'). If lsim_mod is invoked with an output
argument, the simulation is not plotted but the poles returned in the
output argument are ordered according to the $\mathcal{L}_2$-norm. The number of

considered poles is given by the 6th input argument. This way of using `lsim_mod` can be used in order to select the dominant modes. Considering residuals is an alternative approach for the same purpose. For example

```
poldom = plot_res(sysfb,2,1,inf,2);
```

gives the two dominant poles which are -0.6 and -1.3 in conformity with the modal simulation. Here "dominance" is relative to the steady state time response between the second input and the first output. In order to compute the dominant poles of a system (all significant inputs / outputs considered), such computation must be done several times. (In this case, for removing the resulting pole repetition, use the function `clean_ev` with option `'r'`.)

Other examples: see page 133, §3.2.35 and §3.2.40.

1.5.2   Modal controllability

It is very important to have a measure of the controllability and of the observability of each mode. This information is needed before it is decided to move a pole. If some moved poles are weakly controllable or weakly observable the results will not be realistic as leading to irrelevant high gains and to high sensitivity.

1.5.2.1   Classical controllability and observability

The concept of input / output controllability is much more interesting than the classical concept of "controllability", therefore, when we refer to "controllability" in this manual (except in the theoretical part presented in appendix), it should be understood that it is "input / output controllability" that is considered.

Let us recall the classical controllability rank test (see Appendix 4, page 247). The eigenvalue λ_i is not controllable if there exists a non-zero vector u_i such that

$$u_i[A - \lambda_i I \ B] = 0$$

Therefore it is natural to measure the degree of controllability of λ_i as the angle between u_i^* and B. Exact non-controllability corresponds to 90 deg. Therefore the cosine of the angle between the left eigenvector u_i and a column of the matrix B, say the kth one (B_k) is a good measure of the controllability of the pole λ_i from the kth input. So, let us define

$$\text{Degree of (input) controllability of } \lambda_i = \frac{u_i B_k}{\|u_i\|\|B_k\|} \qquad (1.109)$$

This definition is extended to the case where all the columns of B are considered:

$$\text{Degree of (input) controllability of } \lambda_i = \frac{\|u_i B\|}{\|u_i\|\|B\|}$$

By duality, observability can be defined as:

$$\text{Degree of observability of } \lambda_i = \frac{\|C v_i\|}{\|v_i\|\|C\|}$$

The main disadvantage of these measures comes from the fact that angles are related to the state space basis. For a well chosen basis (all states are physically meaningful), the above measures are relevant. However, when identified or interconnected systems are considered it happens that some states have no significance, in this case, this measure is not reliable.

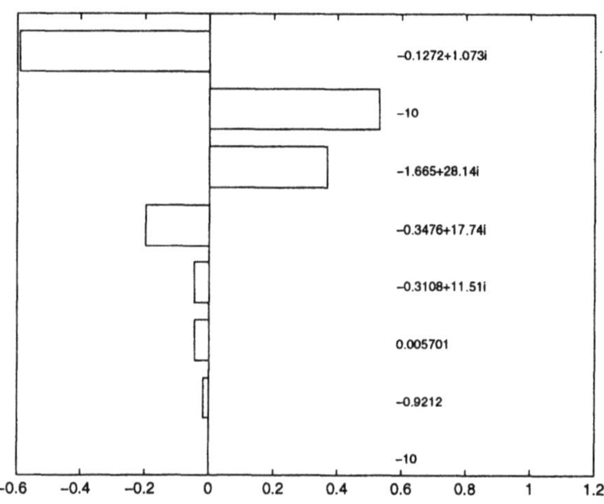

Figure 1.19.   Example of an input/output controllability analysis

1.5.2.2   Input / output controllability

In order to avoid the problems induced by the choice of the state space basis it is better to consider an input / output controllability measure. Furthermore, as we are dealing with modal control, a concept of controllability related to the propensity of poles to be moved by output feedback is of great interest. Consider a closed-loop system denoted (A, B, C, D) ($\rightarrow w_i = 0$ and $t_i = 0$). We have shown that (see (1.99))

$$\Delta \lambda_i = u_i B \Delta K C v_i \qquad (1.110)$$

It is tempting to define a degree of controllability which represents

$$\frac{\Delta|\lambda_i|}{\|\Delta K\|}$$

From Equation (1.110),

$$\Delta|\lambda_i| \leq \|u_i B\|\|\Delta K\|\|C v_i\|$$

of

$$\frac{\Delta|\lambda_i|}{\|\Delta K\|} \leq \|u_i B\|\|C v_i\|$$

Therefore, it is justified to consider the following definition:

Degree of "input / output controllability" of $\lambda_i = \|u_i B\|\|C v_i\|$

See Figure 1.19 for an example.

Comments.

1 - The above degree of "input / output controllability" might be normalized by dividing it by the real part of the poles. Such a normalization would be justified taking into account the fact that the size of pole variation has not the same significance for poles with large negative real part and for poles which are close to the limit of stability. However such a definition would lead to singularity for poles on the imaginary axis.

2 - All measures of the degree of controllability must be considered with caution. Only large differences are significant for comparison of the controllability of two poles.

3 - The proposed measures are based on local properties. An alternative analysis tool consists of plotting the poles for a set of random gains. Then, weak controllability corresponds to dense clouds of closed-loop poles.

Software

Concerned MATLAB-**function**: plot_con. This function permits us to compute the three measures defined in this subsection. These measures can be computed between any subsets of inputs and outputs.

Examples: Observability measures from 2nd and 4th outputs.

```
figure
plot_con(sys,[],[2;4])
[pol,measure] = plot_con(sys,[],[2;4])
```

Input / output controllability measures from 1st and 3rd inputs to 2nd and 4th outputs

```
[pol,measure] = plot_con(sys,[1;3],[2;4])
```
Other example: See the on-line `help2` message of `plot_con`.

1.5.3   Zeros and "almost zeros"

For decoupling it is often necessary to know the zeros of a system (see §2.3). There are several definitions of zeros ([MacFarlane and Karcanias, 1976]), invariant zeros are the values of s at which the rank of the system matrix

$$P(s) = \left[\begin{array}{cc} A - sI & B \\ C & D \end{array} \right]$$

degenerates. In other words, ignoring the special cases where $P(s)$ is rank deficient for all value of s, s_0 is a zero if the minimum singular value of $P(s_0)$ is equal to zero. A numerical technique based on this definition is proposed in [Laub and Moore, 1978]. This toolbox proposes an alternative numerical technique based on optimization rather than on algebra. We will call "almost zeros", the values of s at which the minimum singular value $\underline{\sigma}(P(s))$ admits a local minimum provided that the value of the minimum is small enough. When the minimum is zero, s is an (exact) invariant zero. Finding the significant local minima is not troublesome because the iterative search is efficiently initialized by the "finite" closed-loop poles corresponding to high feedback gains.

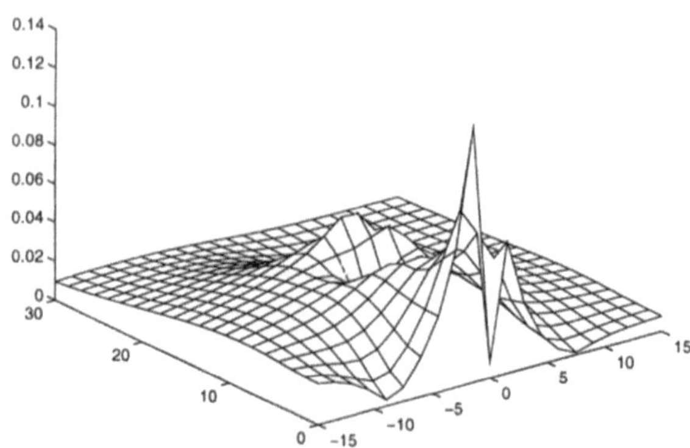

Figure 1.20.   Example of a 3-D plot of the minimum singular value

This information is very important, for example when decoupling cannot be performed exactly. Assume that we want to solve the following equation in the mean square sense

$$\left[\begin{array}{cc} A - \lambda_i I & B \\ E & 0 \end{array}\right]\left[\begin{array}{c} v_i \\ w_i \end{array}\right] \approx 0$$

the choice of λ_i as an almost zero of the triple (A, B, E) will improve decoupling because the size of $\|Ev_i\|/\|v_i\|$ is smaller than for other choices of λ_i.

Similar ideas can be used also for pole assignment by output feedback, see [Magni et al., 1991] for details. Briefly, in this reference, when it is decided to assign a set of $q < p$ right eigenvectors (which form a matrix denoted V_q), the significant almost zeros of the triple (A, V_q, C) must be assigned by the remaining degrees of freedom (assignment of $p - q$ additional eigenvectors), otherwise the assignment is not realistic as leading to high gains.

Software

Concerned MATLAB-**functions**: plot_zer and azer. The first function permits to obtain a 3-D view of the minimum singular value. Figure 1.20 is generated as follows:

```
sys = demodata(3);
plot_zer(sys,[-15:1.5:15],[0:1.5:30]);
```

The computation of the almost zero is done by finding local minima. The function azer is used as follows:

```
[almzer,jsvd] = azer(sys);
```

Seven almost zeros are found: $4.82 \pm 22.43i$, $-1.18 \pm 10.66i$, 7.35, -8.50, -0.046. The corresponding minimum singular values are very low. These almost zeros are in fact true zeros that cannot be found with standard functions (tzero) on account of the approximations introduced by a four digits precision of the matrices A, B, C and D.

The "almost zero direction" v_i can be found using defin_vw. For example, concerning the second almost zero, we have:

```
[v,w] = defin_vw(sys,azer(2),'n',[]);   Note that in the com-
```

plex case, the almost zero direction are given by v(:,1)+i*v(:,2).

1.5.4   Miscellaneous tools

This subsection considers miscellaneous tools for modal analysis. First is mentioned a function that permits the designer to compute the matrices V, W, U, T of page 15 (in the dynamic feedback case, the eigenstruc-

ture concept is revisited as in the comment of page 65). This function is
`eig_fb`, see §3.2.26 for more details.

It is also very useful to treat vectors of eigenvalues. For that purpose,
five functions are available:

- `sort_ev`: this function extracts the entries of a vector that are located in a given area of the complex plane. This area is defined by
  inequalities relative to real, imaginary part and damping ratio. For
  example `pol2 = sort_ev(pol1,'real','>',0,'&','imag','>',0)`,
  → `pol2` is the subset of the unstable entries of `pol1` with positive
  imaginary part. Mostly useful for identifying the eigenvalues that
  must be tuned. See §3.2.48 for more details.

- `vicin_ev`: this function extracts the entries of a vector that are located in small circles around given values. See §3.2.52 for more details.

- `contr_ev`: this function extracts the entries of a vector (poles of a
  given system) that have a given minimum degree of controllability.
  Mostly useful for avoiding tuning of weakly controllable eigenvalues.
  See §3.2.8 for more details.

- `clean_ev`: removes conjugates if a series of complex values (eigenvalues for example) or, on the contrary, add the missing conjugates. Can
  also be used to remove duplication of values in a vector. See §3.2.6
  for more details.

- `choi_ev`: compares two set of complex conjugate values trying to
  recognize, in the second set, the values that are given in the first one.
  This function is mostly called by other functions of the toolbox when
  it is necessary to identify the entry of a given spectrum the closest
  to a value entered (with some approximation) by the user. See §3.2.5
  for more details.

Notes

1 The denomination "input direction" will be justified later. On page 19, it is shown that $u = Ky$ can be written $u = \sum w_i \xi_i$ in which ξ_i are the modes. So the vectors w_i describe the "direction" of the modes as entering in the feedback signal. By duality we will speak of "output directions" for the vectors t_i defined later.

2 If λ_i is not controllable, $V(\lambda_i) \in \mathbb{C}^{n \times m'}$ where $m' > m$ and if λ_i is not observable, $U(\lambda_i) \in \mathbb{C}^{p' \times n}$ where $p' > p$, see page 247.

3 In order to avoid ambiguities in the notations concerning U and $U(\lambda)$ or V and $V(\lambda)$ the parameter λ will not be omitted for the second class of matrices.

4 The normal rank is the rank for almost all values of s. For decoupling, the normal rank should be maximal, *i.e.*, equal to $n + m$.

5 The former form of H must be normalized so that the steady state gain between z_{Ri} and z_i is equal to the unity. Zero-cross coupling in steady state cannot be insured but transient cross-coupling might be better.

6 Lemma 1.2.8 shows that when the feedback loop is closed using observed signals, the vectors u_i can be viewed as an assigned left eigenvectors.

7 When a feedback law is used as depicted in Figure 1.10, $T = 0$ means that a dynamic feedforward structure instead of a feedback structure is obtained. More generally, when projections are used ($T \approx 0$), by a continuity argument, the feedforward structure remains "dominant" which is a natural way to address robustness for the open-loop eigenvalues that are assumed to be well located.

Chapter 2

SOME CONTROL DESIGN PROBLEMS

Contents. Before the presentation of control design problems, several recommendations are listed. A general control design procedure is also proposed. Addressed design problems:

- Pole assignment .... page 117

- Decoupling .... page 129

- Multi-model eigenstructure assignment .... page 139

- Flexible systems control .... page 149

- Structured gain design and controller order reduction ..... page 159

2.1 Recommendations and proposed design cycle

This section summarizes recommendations and gives some general guidelines for modal control design. It is also useful for diagnosis in case of problems.

2.1.1 General recommendations

Note that some illustrative examples given in this manual or in the on-line help messages are just given for illustrating specific points and should not be understood as a good practice of modal control (*e.g.*, the on-line help messages of `fb_tun` in which two random systems are stabilized simultaneously with the same gain).

- *Use of standalone functions* (*e.g.*, `dfb_proj` and `dfb_ins`). This is not the "optimal" way for modal control design because all systems are specific. Note that:

  - The function `dfb_proj` computes a dynamic controller considering projections of open-loop eigenvectors. This function will fail if the open-loop system has multiple poles or non-significant eigenvectors (*e.g.*, eigenvectors corresponding to poles at the origin introduced for integral control).

  - The function `dfb_ins` assigns *all the poles* precisely. This strategy is not relevant if there are non-dominant poles. It must absolutely be avoided if there are weakly controllable poles, unless these poles are precisely "re-assigned" (see below).

  The examples of pages 121 and 123 are good alternative approaches that are compatible with most single-model design problems.

- *Weakly controllable poles.* Unlike with optimal techniques, with modal approaches, weakly controllable poles must be identified so that they are not assigned elsewhere. The input/output controllability degree of each pole can be analyzed using `plot_con` (note that only very large differences between degrees of controllability are significant).

- *Dominant poles.* Roughly, dominant poles are poles with a not insignificant degree of input/output controllability, provided that they are close enough to the origin. Although this concept is somewhat subjective, knowledge of dominant poles is essential for modal control. Usually, all these poles (and only these poles) must be treated. If a feedback law is designed ignoring one of them, such a pole is well controllable and in the controller bandwidth and so, it is likely that it will be sensitive to closure of the feedback loop. There is, therefore, no guarantee that it will be (indirectly) well assigned.

- In most systems, dominant poles are known *a priori*. Usually, poles belonging to the following classes are not dominant: poles of filters, of actuators, stable poles far from the imaginary axis, poles close to the imaginary axis but far outside the controller bandwidth, poles introduced by the controller denominators.

- If no *a priori* knowledge is available, dominant poles can be identified as corresponding to the dominant components of the modal decomposition of (1.107), page 96 (ignoring the modes that cancel each other).

- *Pole motion.* In principle, when a pole is assigned at some given location, it is not relevant to wonder about which open-loop pole has moved to this location. Pole assignment by projection of right eigenvector has a "magic effect[1]" which gives (to some extent) a practical significance to "from open-loop to closed-loop pole motion". The open-loop pole that is "moved" must have a good input/output controllability degree and the distance between open and closed-loop locations must not be too large.

- *Use of observers.* Usually observer and feedback designs are performed independently. With modal control design, as proposed in this toolbox (use of **fb_prop**), it is better to first design the observer, to connect it to the system (**add_obs**), and then to design the feedback (the observer poles are non-controllable, so, non-dominant). This strategy permits the designer to reduce the observer order to the minimum required for assignment of all dominant poles.

- *Proceed by "continuity".* If there are no decoupling requirements, proceed by "continuity" as much as possible by assigning projections of eigenvectors and by shifting "slowly" the poles. If open-loop poles are well located re-assign them at the same location with the same eigenvector. This treatment must be reserved for dominant poles. During this step by step motion of dominant poles, analyze the motion of the non-dominant ones and stop when necessary. If there are more than p (number of outputs) dominant poles to be treated, consider a dynamic feedback (for example an observer which adds a number of observed signals equal to the number of missing measurements).

- *Global constraint on pole assignment.* It is worth keeping in mind that when $D = CB = 0$ (a very common case when actuators are modeled), the sum of all the closed-loop poles remains unaltered when a proportional feedback is applied ($\text{trace}(A + BKC) = \text{trace}(A) + \text{trace}(KCB) = \text{trace}(A)$). Therefore, when dominant poles are moved to the left, usually non-dominant poles (non-dominant in that they are "far" from the imaginary axis) move by the same amount to the right.

- *Repeated eigenvalues.* The tools proposed in this toolbox are not well adapted to the case of repeated eigenvalues. One of the main reasons for this is that the standard MATLAB function `eig` often fails as it returns a singular eigenvector matrix. So, when the considered system has repeated open-loop eigenvalues (several integrators for example) do not use projection of these open-loop eigenvectors (so, do not use the functions `sfb_proj`, `dfb_proj` and `sob_proj`). It is also recommended to assign non-repeated closed-loop poles (for example, if it is expected to use `lsim_mod` for analysis).

2.1.2   Multi-model sub-toolbox

The recommendations given above for single-model design hold in the multi-model case (use of `fb_dyn`). In addition:

- *Coherency in multi-model assignment.* In the single-model case, the problem of assigning twice the same eigenvector (leading to infinite gain) is naturally avoided. In the multi-model case it is sometimes tempting to assign, with similar constraints (as decoupling or projection), two eigenvectors corresponding to two poles of the same nature (for example in aeronautics, Dutch roll mode, short period oscillation mode) of two distinct models. If these models are far from each other, no problem will be encountered, otherwise, by a continuity argument with the single-model case, the feedback will not be realistic.

- *Proceed by "continuity".* New models for design must be added sequentially when analysis detects a new worst case. In this case, the badly located poles of the new worst case model must be moved step by step (projections) starting from the available feedback. When a new assignment constraint is added, it is often necessary to remove a similar one which was already considered for an other model (coherency).

- *Do not consider simultaneously first order and global constraints.* Two functions define constraints for multi-model control design (`eig_cstr` and `dp_cstr`). The first one is used for global eigenstructure assignment, the second one is based on first order approximations for shifting poles. The global constraints might induce large gain variations that are not compatible with first order approximations. It is recommended to close the loop designed using `eig_cstr` before using `dp_cstr`.

- *Use low complexity controllers.* The computation of the criterion is badly conditioned (see page 254), so, the upper limit for the number of poles in the feedback denominators is five. For roll-off properties, add the relevant filters at the plant inputs instead of considering a gain with high degree deficiency between numerators and denominators.

- *Consider feasible design objectives.* Unfortunately we do not provide analysis tools for checking feasibility. Only common sense prevents us from trying to use the multi-model design strategy, for example, in order to find a single controller for two systems of different natures. But sometimes, common sense is not sufficient.

2.1.3 Tuning sub-toolbox

Most of the recommendations given above, for single-model and multi-model design, hold when a feedback is tuned (use of `fb_tun`). In addition:

- *Use a good initial design before tuning.* The proposed techniques are very elementary (with respect to sophisticated optimization techniques) but have nice convergence properties, especially when the product mp is large. Unfortunately there is no guarantee of convergence. Difficulty in convergence increases with the number of models considered together, and above all, with the order of the dynamic controller (the specificity of dynamic controllers is not taken into account in this version of the toolbox).

- *Initialization for tuning a dynamic feedback.* If the dynamic extension of §1.2.6 (page 45) is considered, it is not possible to initialize tuning by a feedback gain of the form:

$$\begin{bmatrix} K_{11} & 0 \\ 0 & 0 \end{bmatrix}$$

see page 86 for more details.

- *Observer-based feedback tuning.* If an observer-based dynamic feedback is tuned, it is not possible to move observer poles (uncontrollable). It is possible to use `contr_ev` to prevent attempts to move these poles. More simply, from the Separation Principle (page 57), it is better to tune directly the equivalent system of Equation (1.78) (this system is the second output argument of `add_obs`).

- *Do not consider simultaneously first order and global constraints.* Most tools for generating constraints are based on first order approximations except `cstr_eig` and `cstr_k`. As discussed in §2.1.2, some care must be exercised as the global constraints might induce gain variations that are not compatible with first order approximations.

2.1.4 Proposed design cycle for direct feedback design

For feedback design, first analyze the system, then, design a nominal feedback. When the nominal feedback is satisfactory, improve it by using multi-model techniques.

2.1.4.1 Models and analysis

Analysis tools necessary to assess design specifications are assumed to be available.

It is also assumed that a representative bank of models is known if robustness problem are to be dealt with. The models should be augmented with the relevant filters *e.g.*, delay models in order to ensure some stability margin (that is not directly taken into account using the multi-model approaches). Relevant integrators should also be added to the models.

Before designing the nominal feedback, it is necessary to know the dominant open-loop modes. Usually, this is *a priori* knowledge. Otherwise, identify these modes by using modal simulations (see §1.5.1) or controllability analysis (see §1.5.2). The functions for dominant mode analysis are `lsim_mod`, `plot_res` and `plot_con`.

2.1.4.2 Nominal design

Multi-model design is not feasible in a single step. It is much better to first design a nominal feedback. The technique used for the nominal design depends mostly on the relative values of the number of dominant modes (say q) and of the number of measurements $(= p)$. For items 2 and 3 below: item 2 leads to a simple design technique, item 3 leads to a more complex but more powerful design technique.

1 - $q \leq p$. Use the technique that is summarized in §1.2.1 (see the recommendations therein). The function for that purpose is `fb_prop`. It might be necessary to go to item 3 below, for example if frequency domain constraints (see `add_cstr`) are required.

2 - $q > p$ *and use of an observer.* The order of the observer should be equal to $q - p$. So, use the technique that is summarized in §1.2.7 (see the recommendations therein). The functions for that purpose are

ob_ins (insensitivity) or **ob_gene** (decoupling and projections). Then, define the system plus observer using **add_obs** and use the technique of §1.2.1 for designing an output feedback, note that the systems has p true outputs plus $q - p$ observed signals, so assignment becomes possible. The functions for assignment are **fb_prop** or **sfb_ins**.

3 - $q > p$ and use of a dynamic feedback in transfer function matrix form. Define the denominators of the transfer function matrix of Equation (1.85) for example, start with a first order denominator. Increase this order only if there are not enough degrees of freedom in the design step. The function to be used for that is **str_cstr**. For feedback design, proceed by continuity (see §2.1). Use **dp_cstr** for small pole shifting, or **eig_cstr** for larger changes.

When the nominal feedback satisfies the specifications, try the assessment tools considering the aforementioned bank of models. If it is okay for all models, the design cycle stops here. Otherwise, note the worst cases and consider multi-model design.

2.1.4.3   Multi-model design

Multi-model design can be performed by using the iterative technique of §1.4 or by using the non-iterative[2] technique of the multi-model sub-toolbox (§1.3). The iterative technique is *a priori* better adapted to improve a nominal feedback design arising from items 1 or 2 of §2.1.4.2. The non-iterative technique is well adapted for improvement of a nominal feedback design arising from item 3 (but also sometimes from item 1).

1 - Iterative approach (tuning). Often it reduces to combining the functions **cstr_qud**, **crit_k** and **fb_tun** in order to push the poles of all worst case models towards the interior of one or more quadrilaterals. Some new worst cases might arise after this tuning process. In this case, try new iterations including these new cases.

2 - Non-iterative approach. Unless the generalized phase control approach is used (only inequality constraints) (§1.3.5), there is a risk of defining redundant constraints (therefore, incompatible, see page 111).

- Any time a new constraint relative to a worst case model is added, as a first step, remove the equivalent constraint relative to the nominal model.

- If it is not possible to guess intuitively which constraints are equivalent between the nominal and worst cases, it probably means that the new constraint is not redundant with the previous ones, so it can be added.

- In some intermediate cases, it might be useful to consider two constraints with a certain degree of redundancy. In order to minimize incompatibility risks, it might be wise to consider two worst cases as "far" as possible from each other, instead of the nominal case and a worst case.

For these steps, the relevant functions are `fb_dyn`, `dp_cstr`, `eig_cstr` and the other functions of the multi-model sub-toolbox.

The proposed multi-model design procedure is in fact a loop (requiring usually two or three iterations) that also includes *multi-model analysis*, possibly exhibiting new worst cases that were not detected after the nominal design. As shown in [Magni et al., 1998], μ-analysis can be very helpful in this design cycle.

2.1.5 Improvement of an existing feedback law

Improvement of an existing feedback is usually straightforward. It consists of re-designing the given feedback as in item 3 of the "nominal design" paragraph presented above. The main difference is that eigenvectors are automatically selected from the ones assigned by the existing feedback. The design steps are:

2.1.5.1 Analysis

Proceed as in §2.1.4, but replace open-loop by closed-loop dominant pole analysis.

2.1.5.2 Nominal design

1 - The first problem consists of selecting the denominator poles of (1.85). Start with a single real pole, if there are not enough degrees of freedom consider additional poles. These poles can be selected from the poles of the initial feedback gain (if it is relevant, take into account in this selection: integral effect, low damped mode control or filtering properties like bandstop filters). Usually this selection is very simple as it reduces to selecting the poles that are the closest to the imaginary axis.

2 - Using the multi-model sub-toolbox tools (for example `fb_dyn`, `dp_cstr`, `eig_cstr`) assign the eigenvectors that were considered as dominant at the previous analysis step.

2.1.5.3 Improvement

When an equivalent modal controller is derived, it can be improved as in item 2 of the "multi-model design" paragraph presented above.

2.2 Single-model pole placement

Contents. This section addresses the problem of eigenstructure assignment when there is no decoupling objectives, so no *a priori* desirable choice of the right eigenvectors.

- From open-loop to closed-loop .... §2.2.1

- Insensitivity .... §2.2.2

- Tuning .... §2.2.3

The first two techniques are specific to single-model design. Pole placement is also dealt with in the multi-model design section (§2.4).

References. The strategy of §2.2.1 was first introduced in the SISO case by de Larminat, see for example [De Larminat and Houisot, 1993]. The proposed generalization to MIMO systems is based on the same idea: settling time and damping ratio requirements are met with minimum control and minimum undesirable "side effects" if the deviation between closed-loop and open-loop behaviors is minimized (assignment of open-loop eigenvector projections). The strategy for insensitivity of §1.2.3 - §2.2.2 was first introduced in [Moore and Klein, 1976].

2.2.1   From open-loop to closed-loop

It is often necessary (especially in aeronautics) to preserve, in closed-loop, the open-loop behavior. As right eigenvectors distribute the modes to the states and outputs, a natural choice consists of assigning right eigenvectors as being as close as possible as they are in open-loop. For that purpose we can use the projection technique of page 22. From experience, such a strategy leads to good robustness. Intuitively, such a design is the one with minimum "side effects" when the poles are shifted. Therefore robustness is naturally expected to be good. An alternative justification of this remark is proposed now.

2.2.1.1   Pole dispersion

Usually open-loop systems, even if they are unstable or have bad stability properties, have eigenvalues which do not vary too much with respect to operating point variations. If right eigenvectors are preserved, in turn, the left eigenvectors are also more or less preserved. Therefore, the quantities $u_i \Delta A v_i$, which quantify pole dispersion, (see Lemma 1.2.2) are also preserved. Using such a strategy, when an open-loop pole is shifted to the left, it is expected that the clouds (corresponding to operating point variations) of open-loop poles are also globally shifted to the left but are not spread over a larger area.

In order to illustrate this discussion, let us consider the lateral RCAM control design problem (see §5.1.2). The open-loop pole map is given in Figure 2.1.

```
rcampole('lat',zeros(2,4),1)
```

There are three clouds of poles, one is outside the real axis, both others lie on it. Let us consider the lateral nominal model:

```
sys = rcamdata('lat',0,1);
```

The open-loop nominal eigenvalues are $\{-0.23 \pm 0.59i, -0.18, -1.3\}$ (plus the actuator eigenvalues $\{-6.66, -3.33\}$ that have a similar degree of controllability (type `plot_con(sys,[1:2],[1:4])`) but are far from the origin, therefore can be ignored). We are going to move

- $-0.23 \pm 0.59i$ to $-0.6 + 0.6i$. The corresponding assigned eigenvector is the projection of the open-loop eigenvector associated with $-0.23 \pm 0.59i$.

- -0.18 to -0.6. The corresponding assigned eigenvector is the projection of the open-loop eigenvector associated with -0.18.

- -1.3 and the corresponding open-loop eigenvector are preserved.

```
pol     = [-0.60+0.60*i -0.60 -1.3];
def_pb = [-0.23+0.59*i -0.18 -1.3];
k = fb_prop(sys,0,pol,'p',def_pb);
rcampole('lat',k,1)
```

The resulting pole map is given in Figure 2.2. The pole cloud outside the real axis, although a little expanded, seems to be shifted to the left as a whole.

2.2.1.2   A general strategy for state feedback

The design strategy discussed above is made systematically with the function sfb_proj in the state feedback case (generalized later in the observer-based dynamic output feedback case, see dfb_proj). The user must choose a settling time and a damping ratio[3]. Both design parameters define an area of the complex plane.

- open-loop poles belonging to this area are re-assigned at the same location with the same eigenvectors.

- open-loop poles outside the area are shifted to the area boundary and the assigned right eigenvectors are selected by projection.

- weakly controllable poles are ignored.

For example, settling time 5 sec and damping ratio larger than 0.7:

```
sys = rcamdata('lat',0,1);
k = sfb_proj(sys,5,0.7,'p');
eig_fb(sys,k,'s')
```

Note that, in order to avoid repeated pole assignment[4], when some eigenvalues are shifted towards the boundary, the function sfb_proj performs assignments at random in a narrow neighborhood of the boundary. This fact explains the small differences when k is computed several times as above.

2.2.1.3   Extension to observer-based dynamic feedback

The use of an observer with the specific "Kalman filter structure" (see page 50) permits us to dualize the above procedure. Now, it is open-loop *left* eigenvectors that are projected. On page 51 is discussed the

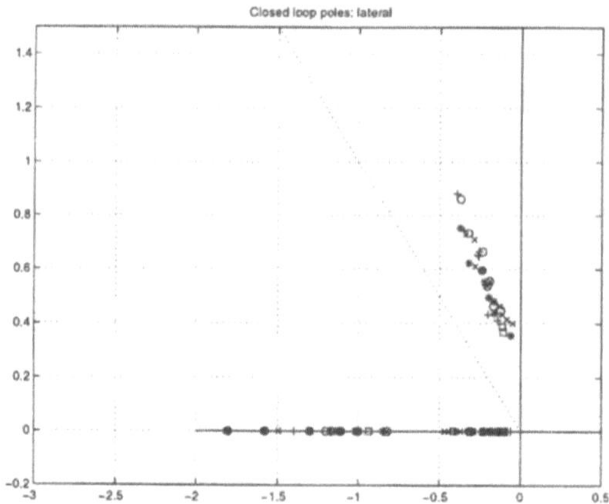

Figure 2.1.   RCAM open-loop pole map

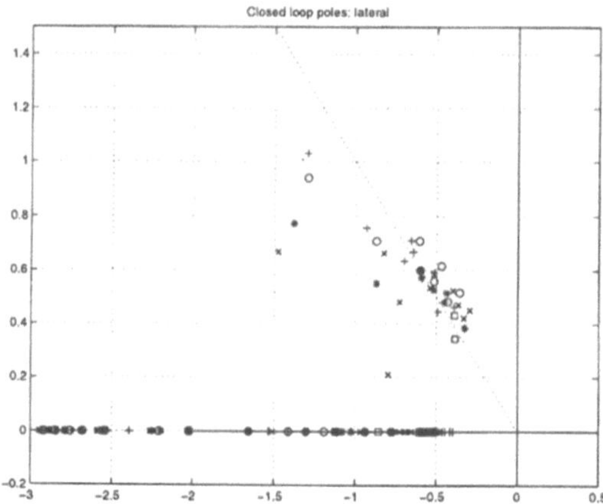

Figure 2.2.   RCAM closed-loop pole map: design by projection of open-loop eigen-vectors

fact that this strategy leads to "minimum side effects". Furthermore, the discussion on pole dispersion can be dualized.

The function dfb_proj performs this design strategy. For example, for obtaining a settling time of 5 seconds and a minimum damping ratio equal to 0.7:

```
sys = rcamdata('lat',0,1);
fb = dfb_proj(sys,5,0.7,'p');
eig_fb(sys,fb)
```

The function **dfb_proj** leads to an n-order dynamic feedback. It is possible to build an observer with reduced order, but such a design must be adapted to the system dealt with: The designer must build the controller step by step as follows.

- Design an observer using the function **ob_gene**. The order can be chosen as the difference between the number of dominant poles and the number of measurements.

- Connect this observer to the system using **add_obs**.

- Using **fb_prop**, compute an output feedback for the system plus observer.

- Compute the equivalent form of the observer-based dynamic feedback using **obs2dfb**.

Example: Let us consider a system with two outputs and four dominant poles

```
sys = [1 0 0 0;0 0 0 1]*demodata(1);
```

A 2nd order observer (projection of one complex left eigenvector) is designed

```
polo  = -1+i;
polo0 = -0.12+1.07i;
[U,T,Pi] = ob_gene(sys,polo,'p',polo0);
```

it is connected to the system

```
syso = add_obs(sys,U,T,1);
```

The proposed feedback design consists of the projection of open-loop right eigenvectors (4 outputs for 4 dominant poles)

```
polc  = [-0.9+0.9i -1 -0.6];
polc0 = [-0.12+1.07i -0.92 0.0057];
k = fb_prop(syso,0,polc,'p',polc0);
```

The corresponding dynamic feedback gain is computed as follows

```
fb = obs2dfb(sys,U,T,1,k);
[eig_fb(sys,fb) eig_fb(syso,k)]
```

2.2.2   Closed-loop eigenvalue insensitivity

The criterion of Equation (1.45) will be minimized:

$$J = \sum_{i=1}^{n} \|u_i\|\|v_i\|$$

The algorithm that is used is detailed on page 35. This method is reported in this book because it is one of the most popular modal design techniques. However, we recommend to use preferably the techniques of the §2.2.1 as these techniques do not involve iterations and do not require a precise definition of all the closed-loop poles.

2.2.2.1   Insensitivity by state feedback

The function for minimizing closed-loop eigenvalue sensitivity is `sfb_ins`, example:

```
sys = rss(8,6,3);
pol = [-5+5*i;-2.5;-0.6+0.6*i;-1.2;-0.8;-0.4];
k = sfb_ins(sys,0,pol,10);
eig_fb(sys,k,'s')
```

2.2.2.2   Insensitivity by output feedback

In the proportional output feedback case (with $p < n$ measurements), only p poles are assigned. Minimizing the sensitivity tends to spread over a larger area the poles which are not fixed, some non fixed poles might become unstable. Nevertheless, if the number of measurements is equal to the number of dominant poles such problems are likely to be avoided. In this case, for computing a proportional output feedback, `sfb_ins` must be invoked with option `'o'`:

```
pol = [-5+5*i;-0.6+0.6*i;-0.8;-0.4];
k = sfb_ins(sys,0,pol,10,'o');
```

In order to obtain more reliable results than using `sfb_ins` with option `'o'`, it is suggested to use the algorithm of §1.4.4, page 92. Let us recall that this algorithm combines the insensitivity criterion reduction together with pole assignment in a trapezium (in order to limit the tendency of non assigned eigenvalues to spread over a larger area).

2.2.2.3   Observer-based insensitive output feedback

The use of an observer with the specific "Kalman filter structure" (see page 50) permits us to dualize the state feedback insensitivity optimization algorithm. The function **dfb_ins** computes the optimal state feedback and optimal observer defined in that way.

```
sys = rss(8,6,3);
polc = [-5+5*i;-2.5;-0.6+0.6*i;-1.2;-0.8;-0.4];
polo = [-5+5*i;-2.5;-0.6+0.6*i;-1.2;-0.8;-0.4];
fb = dfb_ins(sys,polc,polo,10);
eig_fb(sys,fb)
```

The function **dfb_ins** leads to n-order dynamic feedback gains. It is possible to build an observer with reduced order but such a design must be adapted to the system dealt with. The designer must build the controller step by step as follows.

- Design an insensitive observer using the function **ob_ins**. The observer order can be chosen as the difference between the number of dominant poles and the number of measurements (the observer poles are not controllable, so can be ignored).

- Connect this observer to the system using **add_obs**.

- Using **sfb_ins** (with option 'o' for output feedback) compute an insensitive output feedback for the system plus observer.

- Compute the state space form of the observer-based dynamic feedback using **obs2dfb**.

Example: Let us consider a system with two outputs and four dominant poles

```
sys = [1 0 0 0;0 0 0 1]*demodata(1);
```

A 2nd order insensitive observer is designed

```
polo = [-1+i];
[U,T,Pi] = ob_ins(sys,polo);
```

it is connected to the system

```
[syso1,syso2] = add_obs(sys,U,T,1);
```

The proposed feedback is designed by eigenvalue sensitivity minimization (4 outputs for 4 dominant poles)

```
polc = [-0.9+0.9i -1 -0.6];
k = sfb_ins(syso2,0,polc,[],'o');
```

The corresponding dynamic feedback gain is computed as follows

```
fb = obs2dfb(sys,U,T,1,k);
[eig_fb(sys,fb) eig_fb(syso1,k)]
```

2.2.3   Tuning of a dynamic controller

Tuning of a controller will be considered again later in a multi-model setting (page 145). Here, is considered a simple academic problem. The controller is dynamic. As only proportional gains can be directly tuned using the "tuning sub-toolbox", in the dynamic case the tuning procedure must be adapted. First, the use of the dynamic extension of §1.2.6 is illustrated. The same example will also be considered using an observer based approach.

2.2.3.1   Use of a dynamic extension

The proposed tuning procedure is as follows.

- Find an initial *dynamic* gain fb0, for example using an observer-based design technique or using fb_dyn (an important recommendation is given on pages 88 and 112).

- Choose nc the order of the feedback and define dyn (*e.g.*, dyn = zeros(nc,1);).

- Add a dynamic extension to the model to be tuned (see (1.53)): sysdyn = add_dyn(sys,dyn);

- Transform the initial dynamic gain into a proportional one (see (1.55)): K0 = dyn2sta(fb0,dyn);

- Tune K0. The resulting gain is denoted K.

- Transform the final gain K into a dynamic feedback (see (1.54)): fb = sta2dyn(K,dyn);

Example. First, an initial random system is generated, an initial feedback gain fb0 is computed (by dividing by 2 the imaginary part of the eigenvalues with damping ratio less than 0.7). Random initial system:

```
rand('seed',0); randn('seed',0);
sys = rss(6,2,2);
```

Gain structure and criterion

```
CSTR = str_cstr(2,2,-1);
```

```
CRIT = ktf_crit(CSTR,0,i*[0 0.01 0.1 1]);
```

Selection of treated eigenvalues (damping ratio < 0.7)

```
lam0 = eig(sys);
lam1 = sort_ev(lam0,'d','<',0.7);
```

Assignment of eigenvalues with imaginary part divided by 2

```
lam1new = real(lam1)+0.5*i*imag(lam1)
CSTR = eig_cstr(CSTR,sys,lam1new,'p',lam1);
[fb0,K0,dyn] = fb_dyn(CSTR,CRIT);
```

The definition of the vector dyn and transformation of fb0 into a proportional gain K0 are already treated by fb_dyn (K0 = dyn2sta(fb0,dyn)), so it remains to treat the system:

```
sysdyn = add_dyn(sys,dyn);
```

Now, K0 can be tuned. The relevant poles are shifted towards the trapezium with corners $-5 + 5i, -1 + i, -5, -1$.

```
Quad = [-5+5i -1+i;-5 -1];

K = K0;
for ii = 1:100;

    lam = eig_fb(sysdyn,K);
    disp(dist_qud(sysdyn,Quad,lam,K));
    if dist_qud(sysdyn,Quad,lam,K) < 0.2;
        disp('BREAK');
        break;
    end;
    CSTR = cstr_qud([],sysdyn,Quad,lam,0.01,K);

    k = fb_tun(CSTR,[]);
    K = K + k;
end
```

Finally, the gain K is transformed into a dynamic gain relative to the original system sys.

```
fbtun = sta2dyn(K,dyn);
[eig_fb(sysdyn,K) eig_fb(sys,fbtun)]
```

2.2.3.2   Use of an observer

The proposed tuning procedure is as follows.

- Find an initial *dynamic* gain fb0, for example using an observer-based design technique.

- nc denotes the order of the feedback. Transform the initial feedback into a nc-order observer-based proportional feedback K0 using dfb2obs.

- Connect the observer to the system using add_obs.

- Tune K0. Note that observer-based feedback design induces uncontrollable modes (that cannot be tuned). In order to prevent attempts to tune these poles, use the function contr_ev that selects controllable poles. It is also possible (and better) to tune the gain relative to the equivalent system of Equation (1.78) (this system is the second output argument of add_obs), but in order to illustrate the use of the function contr_ev, we chose below the bad solution.

- Transform the final gain K into a standard dynamic feedback using obs2dfb.

Example. In order to illustrate this design procedure, the system **sys** and the initial feedback **fb0** of the previous example are considered. First, fb0 is transformed into an observer-based form and the corresponding observer is connected to the system.

```
[U,T,Pi,Q,K0] = dfb2obs(sys,fb0,0);
sysobs = add_obs(sys,U,T,Q);
```

The tuning loop is similar to the previous one, note that the function contr_ev is added in order to ignore the uncontrollable observer poles.

```
Quad = [-5+5i -1+i;-5 -1];

K = K0;
for ii = 1:100;

    lam = eig_fb(sysobs,K);
    lam = contr_ev(sysobs,K,lam,0.001);
    disp(dist_qud(sysobs,Quad,lam,K));
    if dist_qud(sysobs,Quad,lam,K) < 0.2;
        disp('BREAK'); break;
```

```
      end;
      CSTR = cstr_qud([],sysobs,Quad,lam,0.02,K);

      k = fb_tun(CSTR,[]);
      K = K + k;
   end
```

This feedback gain can be converted to a standard dynamic feedback as
follows.

```
      fbtun = obs2dfb(sys,U,T,Q,K);
```

2.3   Decoupling

Contents. Non-interactive control:

- Considering a lateral aircraft model, the problem of non-interactive control is treated by designing a constant output feedback and a constant feedforward .... §2.3.1

- The same results are recovered by means of a dynamic feedforward (partial inversion) .... §2.3.2

- Similar problem with an observer-based feedback .... §2.3.3

Exact Loop Transfer Recovery .... §2.3.4

References. Non-interactive control has received much attention from a theoretical point of view (Geometric Approach). For most applications, the initial work of [Falb and Wolovich, 1967, Gilbert, 1969] (exact decoupling by state feedback) remains the most relevant approach. However, the decoupling technique of [Falb and Wolovich, 1967, Gilbert, 1969] is based on several rank conditions that need a good understanding of the system for interpretation. An alternative technique based on right eigenstructure assignment is proposed in literature (see for example [Farineau, 1989] or [Liu and Patton, 1998] and the references therein). It is the eigenstructure assignment point of view that is considered here. The Exact Loop Transfer Recovery technique, as illustrated here, comes from [Apkarian et al., 1989, Magni and Mouyon, 1991].

2.3.1 Non-interactive control by proportional feedback

Non-interactive control by eigenstructure assignment is detailed in §1.2.4 (page 37). First, is applied the technique of §1.2.4, it consists of designing a proportional output feedback gain. In order to enlarge the applicability of decoupling, two extensions will be proposed

- Feedforward-based design, see §2.3.2.

- Observer-based design (case where there are less measurements than dominant modes), see §2.3.3.

Let us consider the lateral decoupling problem of the RCAM (model described on page 290). The proposed technique consists of assigning right eigenvectors (transmissions from modes to regulated outputs) and to design a constant feedforward (transmissions from reference inputs to modes). The first regulated output is the angle of side slip

$$z_1 = \beta = C_1 x$$

where C_1 denotes the first row of C and the second one is the roll angle

$$z_2 = \phi = C_4 x$$

where C_4 denotes the fourth row of C.

Four classes of modes are considered in Figure 1.3 (page 39). Here, the considered system is controllable and has one zero between u and (z_1, z_2). The zero value is -13.8, that is far form expected closed-loop dynamics, therefore this zero will be considered as a "zero at infinity", in other words it will be ignored.

Finally, two classes of poles must be considered: those associated with z_1 and those associated with z_2. Figure 2.3 is an adaptation of Figure 1.3 to the considered problem (in which are added the non-dominant modes).

The considered system is generated as follows:

```
sys0 = rcamdata('lat',0,1);
```

The feedback gain is computed such that the right eigenvectors relative to the closed-loop eigenvalues $\lambda_1 = -0.6+0.6j$, $\lambda_2 = -0.6$ and $\lambda_3 = -1.3$ satisfy

$$C_1 v_1 = 0 \text{ and } C_4 v_2 = C_4 v_3 = 0$$

So,

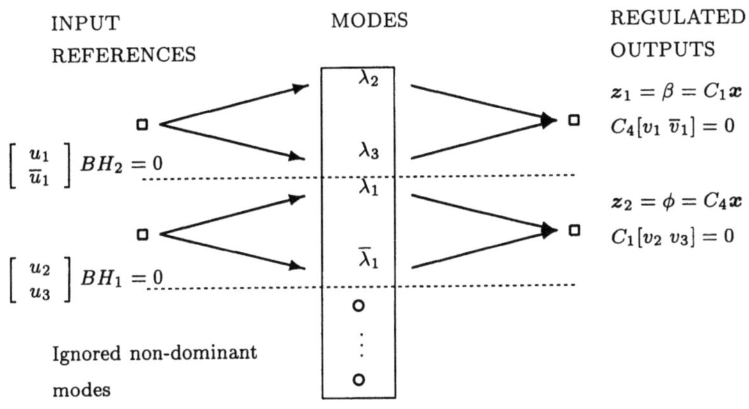

Figure 2.3. Non-interactivity for a lateral aircraft model

```
p_new(1) = -0.6+0.6*i; key(1) = 'n'; def_pb(1) = 1;
p_new(2) = -0.6;       key(2) = 'n'; def_pb(2) = 4;
p_new(3) = -1.3;       key(3) = 'n'; def_pb(3) = 4;
K1=fb_prop(sys0,0,p_new,key,def_pb);
```

Computation of the feedforward gain. Following Figure 2.3, the matrix H can be computed by solving (mean squares) the following equations

$$\left[\begin{array}{c} u_1 \\ \overline{u}_1 \end{array}\right] BH_2 = 0$$

and

$$\left[\begin{array}{c} u_2 \\ u_3 \end{array}\right] BH_1 = 0$$

however, an alternative almost similar technique was proposed in §1.2.4. It consists of computing the feedforward gain that performs decoupling in steady state. This gain can be computed by invoking `ff_stat`.

```
H1 = ff_stat(sys0,K1,[1 4]);
```

Modal simulation are considered in order to check that the expected decoupling properties are satisfied (by invoking the function `lsim_mod`). First, is considered the step responses from the reference input $\beta_R =$

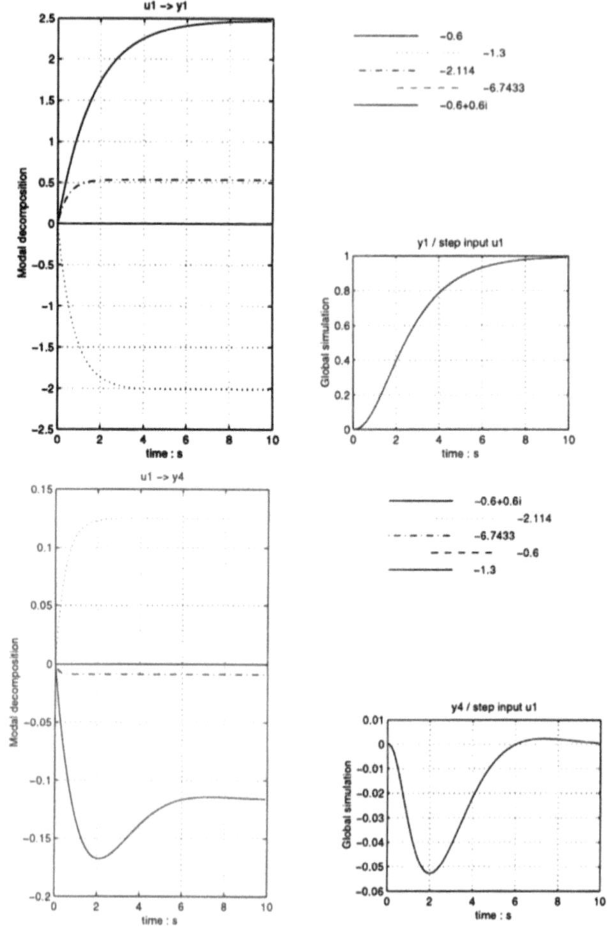

Figure 2.4. Proportional feedback: time responses of β (top) and ϕ (bottom) to a step reference input β_R

z_{1R} to $\beta = z_1$ and to $\phi = z_2$ (Figure 2.4). Then, step responses are considered relative to a reference input $\phi_R = z_{2R}$ (Figure 2.5). The second and third simulations represent cross-couplings. ('y1' $= z_1 = \beta$ and 'y4' $= z_2 = \phi$)

```
sys1 = feedback(sys0,K1,1);
sys1 = sys1*H1;
```

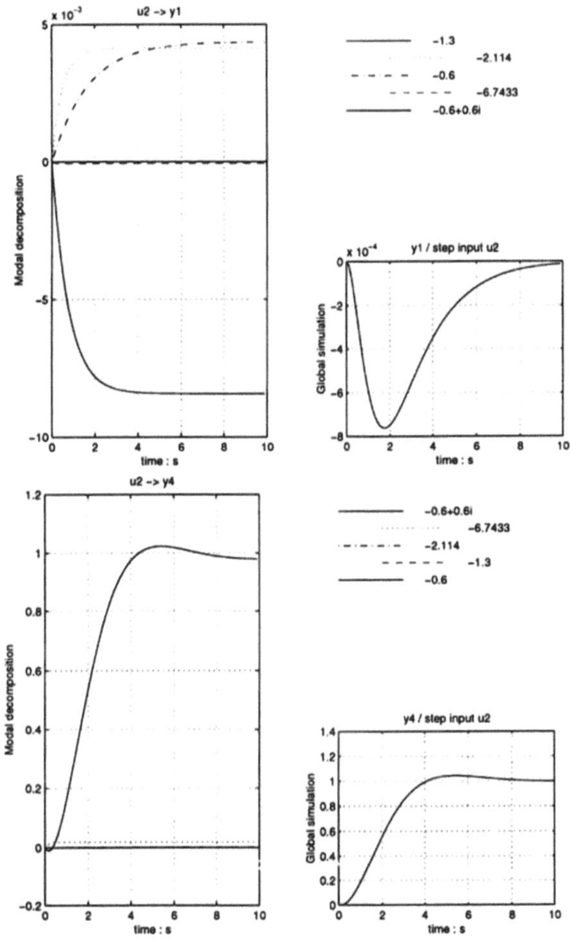

Figure 2.5. Proportional feedback: time responses of β (top) and ϕ (bottom) to a step reference input ϕ_R

```
lsim_mod(sys1,'u1',10,0,'y1')
lsim_mod(sys1,'u1',10,0,'y4')
lsim_mod(sys1,'u2',10,0,'y1')
lsim_mod(sys1,'u2',10,0,'y4')
```

As expected, it appears (see Figures 2.4 and 2.5) that the mode with dynamic $-0.6 \pm 0.6j$ "is not seen" in the ϕ-time response and that the modes $-0.6, -1.3$ "are not seen" in the β-time response.

2.3.2   Non-interactive control by dynamic feedforward

The right eigenvector "assignment" is made by cancellation of some poles (the considered system being stable in open-loop, it is not necessary to stabilize it beforehand). The right eigenvectors corresponding to the poles introduced by the dynamic feedforward are assigned in order to ensure decoupling (same assignment than in the feedback case). The considered system is:

```
sys0 = rcamdata('lat',0,1);
```

Computation of the dynamic feedforward that replaces right eigenvector assignment:

```
p_new(1) = -0.6+0.6*i; key(1) = 'n'; def_pb(1) = 1;
p_new(2) = -0.6;        key(2) = 'n'; def_pb(2) = 4;
p_new(3) = -1.3;        key(3) = 'n'; def_pb(3) = 4;
p_old = [-0.23+0.59*j -0.18 -1.3];
zer = p_old;
F2 = ff_assgn(sys0,0,p_old,zer,p_new,key,def_pb);
```

Computation of the feedforward gain so that the steady state gain becomes the identity matrix:

```
H2 = ff_stat(sys0,0,[1 4]);
```

Simulations:

```
sys2 = sys0*F2*H2;

lsim_mod(sys2,'u1',10,0,'y1')
lsim_mod(sys2,'u1',10,0,'y4')
lsim_mod(sys2,'u2',10,0,'y1')
lsim_mod(sys2,'u2',10,0,'y4')
```

The simulation results of Figures 2.6 and 2.7 are very similar to those obtained in the state feedback case (Figures 2.4 and 2.5).

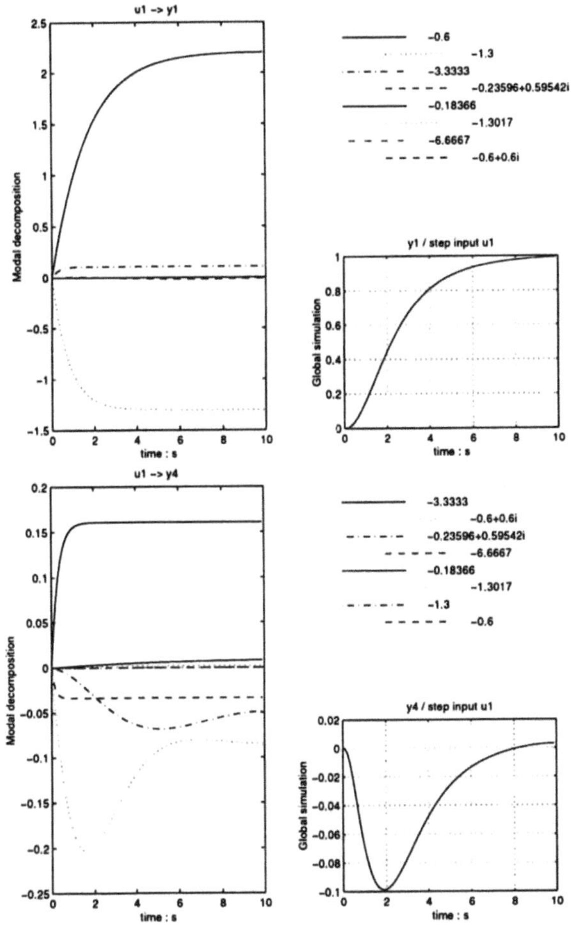

Figure 2.6. Feedforward / cancellation: time responses of β (top) and ϕ (bottom) to a step reference input β_R

2.3.3 Observer-based non-interactive control

Assume that we have only two measurements (β and ϕ) at our disposal. We would like to use a two dimensional observer in order to have four variables for feedback (two measurements and two observed variables) so that the decoupling technique of §2.3.1 can still be used. The

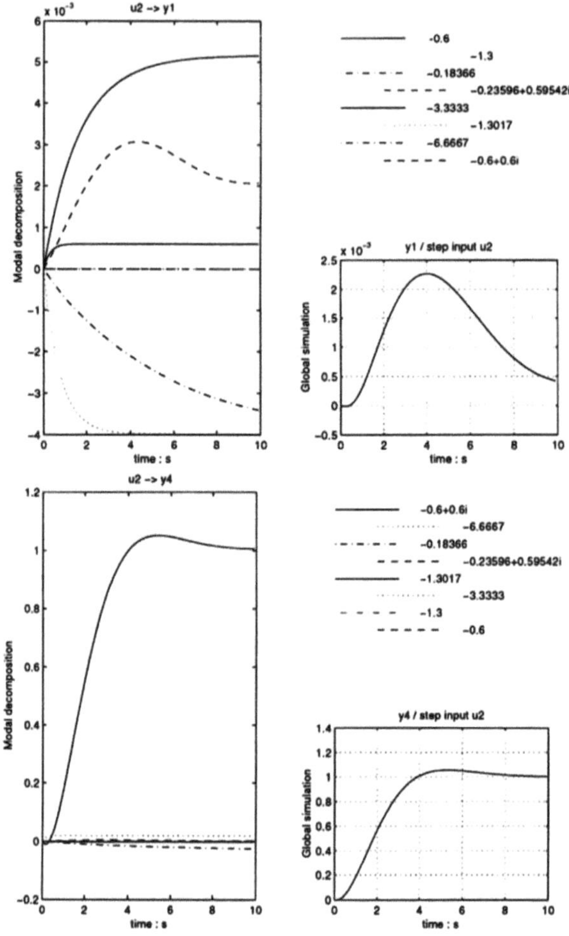

Figure 2.7.   Feedforward / cancellation: time responses of β (top) and ϕ (bottom) to a step reference input ϕ_R

main functions involved here are **ob_gene** (observer design) and **add_obs** (connection of an observer to a system as in Figure 1.10). Model used:

```
sys0 = [1 0 0 0;0 0 0 1]*rcamdata('lat',0,1);
```

An elementary observer with poles $-0.8 \pm 0.8i$ is designed by projection of the open-loop left eigenvector relative to $-0.2360 + 0.5954i$. Then, the observer is connected to the system.

```
[U,T] = ob_gene(sys0,-0.8+0.8i,'p',-0.2360+0.5954i);
syso  = add_obs(sys0,U,T,1);
```

The feedback K3 and a feedforward gain H3 are computed as in §2.3.1.

```
p_new(1) = -0.6+0.6*i; key(1) = 'n'; def_pb(1) = 1;
p_new(2) = -0.6;       key(2) = 'n'; def_pb(2) = 2;
p_new(3) = -1.3;       key(3) = 'n'; def_pb(3) = 2;
K3 = fb_prop(syso,0,p_new,key,def_pb);
H3 = ff_stat(syso,K3,[1 2]);
```

Simulation results are obtained as follows:

```
sys3 = feedback(syso,K3,1);
sys3 = sys3*H3;

lsim_mod(sys3,'u1',10,0,'y1')
lsim_mod(sys3,'u1',10,0,'y2')
lsim_mod(sys3,'u2',10,0,'y1')
lsim_mod(sys3,'u2',10,0,'y2')
```

Decoupling properties are very close to those obtained using four measurements.

Comment. The observer-based structure can be replaced by an equivalent more standard structure with dynamic feedback (FB3) and feedforward (FF3). Theses gains are computed using **obs2dfb**

```
[FB3,FF3] = obs2dfb(sys0,U,T,1,K3);
```

and the equivalent system is

```
sys3 = feedback(sys0,FB3,1)*FF3;
H3 = ff_stat(sys3,0,[1 2]);
sys3 = sys3*H3;
```

2.3.4 Illustration of Exact Loop Transfer Recovery

The ELTR problem is defined on page 60. Here, the *number of inputs* of the considered system *is strictly less than the number of outputs*. The direct transmission matrix "D" is equal to zero. In this case, it can be shown that the state feedback ELTR problem is generically solvable.

```
sys = rss(8,3,2);
[a,b,c,d] = ssdata(sys);
sys = ss(a,b,c,0*d);
```

The observer that solves Equation (1.81) is designed as follows (the ELTR property is insured by the decoupling requirements defined by def_pb that is $u_i B = 0$, see page 60).

```
pol = [-0.5+i*0.5;-1;-1+1*i];
def_pb = [[1;2] [1;2] [1;2]];
[U,T,Pi] = ob_gene(sys,pol,'n',def_pb);
```

The observer (U,T,Pi) is connected to the system. A minimum order *state* observer is used (that justifies the form of the fourth input argument of add_obs below, see page 50).

```
syso = add_obs(sys,U,T,inv([c;U]));
```

In order to check the ELTR property, let us consider an LQR state feedback designed as follows:

```
K = - lqr(a,b,10*eye(8,8),eye(2,2));
```

The ELTR property can be verified by checking that both following transfer function matrices are equal.

```
sys1 = K*ss(a,b,eye(8,8),0);
sys2 = K*syso;
bode(sys1-sys2)
```

As expected, the above Bode plot shows that the difference is less than -200 dB for all frequencies.

Comment. The state feedback ELTR problem is also generically solvable when *the number of inputs is equal to the number of outputs provided that all the zeros have negative real part.* More generally, state feedback loop transfer properties cannot be recovered if at least one zero has positive real part.

2.4 Multi-model eigenstructure assignment

Contents.

- Multi-model design using `fb_prop` .... §2.4.1
- Multi-model design using `fb_dyn` .... §2.4.2
- Multi-model design using `fb_tun` .... §2.4.3

References. Multi-model control is usually treated by parameter optimization, see for example [Joos, 1999, Joos, 1997]. Quadratic stabilization is an alternative way to treat a family of models (continuum), see for example [Garcia et al., 1997] and several LMI-based control design techniques. But treating a continuum of models induces some conservatism (a single Lyapunov function for the continuum). Using multi-model design there is no conservatism, furthermore the solution can be algebraic. Indeed, the techniques using `fb_prop` and `fb_dyn` are based on Least Squares or on Linear Quadratic Programming (convergence guaranteed). These techniques were proposed in [Magni et al., 1997a, Magni, 1999, Magni et al., 1998]. In [Magni et al., 1998], it is shown how μ-analysis can be used in order to improve the robustness potentialities of multi-model design.

A single feedback is designed for all the RCAM flight envelope. This is a too ambitious objective, as from industrial experience, gain scheduling must to be used. This section must be understood as a software demonstration rather than a realistic use of robust (multi-model) eigenstructure assignment.

2.4.1 Multi-model design with a proportional gain

In §2.3.1 was considered a flight control system with a pilot in the loop (two reference inputs). Here, it is an autopilot problem that is considered. The pilot is replaced by two integrators as shown in Figure 5.1 (page 292).

The nominal model is denoted sys0

```
sys0 = rcamdata('lat',0,2,0);
```

Initial analysis. The considered system has clearly six dominant poles which are the four dominant poles of the natural aircraft plus two integrators poles. As there are six measurements, proportional feedback design is feasible.

Initial nominal feedback design. Three assigned eigenvectors are associated with β and three other ones are associated with ϕ (we proceed as in §2.3.1 with two additional poles). For a justification of this strategy and details on the model, see [Magni et al., 1997b].

```
pol0 = [-0.6+0.6*i -0.5 -1.3 -1.0 -0.6];
K0   = fb_prop(sys0,0,pol0,'n',[4 4 1 1 1]);
```

Worst case analysis. When K0 is applied to the bank of available models, the worst case appears to be model number 12 (denoted sys1), corresponding to high mass (see page 294). For this system, a badly damped pole is assigned at -0.1084 + 0.4870i by K0.

```
sys1 = rcamdata('lat',12,2,0);
```

Multi-model feedback design. The badly damped pole of sys1 is moved to -0.5 + 0.5i while the similar constraint relative to sys0 is removed. Removing a constraint is necessary for avoiding redundancy, but the main reason is that, in the proportional output feedback case, there are not enough degrees of freedom for considering an additional constraint. Up to now, we used the function **fb_prop** for proportional feedback design. Here, in order to take into account the fact that all constrains are not relative to a single model, we use a similar function `defin_vw` that just compute the constraints relative to the feedback gain.

```
oldpol0 = pol0(2:5);
newpol0 = pol0(2:5);

oldpol1 = -0.1084+0.4870i;
newpol1 = -0.5+0.5i;

[V0,W0,CV0] = defin_vw(sys0,newpol0,'p',oldpol0,K0);
[V1,W1,CV1] = defin_vw(sys1,newpol1,'p',oldpol1,K0);
```

It remains to apply Equation (1.32) (page 28):

```
K1 = [W0 W1]*inv([CV0 CV1]);
```

Analysis of the results.

```
[eig_fb(sys0,K1) eig_fb(sys1,K1)]
```

The closed-loop eigenvalues are given below, we recognize `newpol0` in the first column and `newpol1` in the second one.

```
                -5.0563              -5.8458
                -1.4421 + 1.7745i    -1.7911
                -1.4421 - 1.7745i    -0.7681 + 0.6551i
newpol0 -> -1.3000                   -0.7681 - 0.6551i
newpol0 -> -1.0000                   -0.5000 + 0.5000i <- newpol1
newpol0 -> -0.5000                   -0.5000 - 0.5000i
                -0.6168              -0.5985 + 0.2760i
newpol0 -> -0.6000                   -0.5985 - 0.2760i
```

Poles and step responses are analyzed:

```
rcampole('lat',K1,2,0)
axis([-3 0.5 -0.5 3])

rcamstep('lat',K1,eye(2,2),2,0)
```

The set of poles assigned for all available models of the RCAM is given in Figure 2.8, the corresponding step responses are given in Figure 2.9. The badly damped models appearing in Figure 2.8 correspond to low mass. This result is logical because the proposed design was relative to nominal and high mass (the correspondences between models and symbols as used in Figure 2.8 are given on page 294).

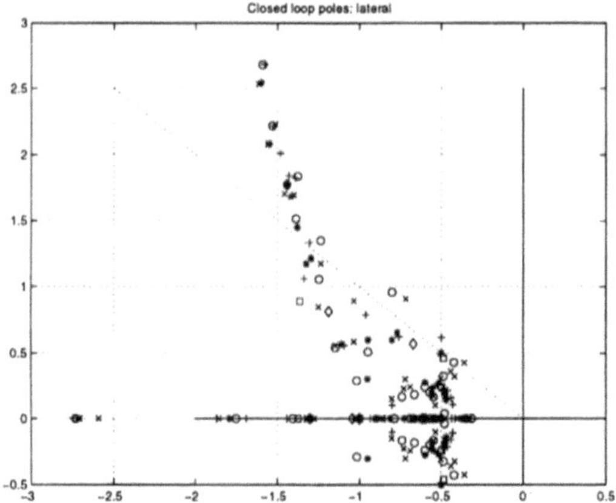

Figure 2.8. Proportional multi-model feedback, poles of all RCAM models.

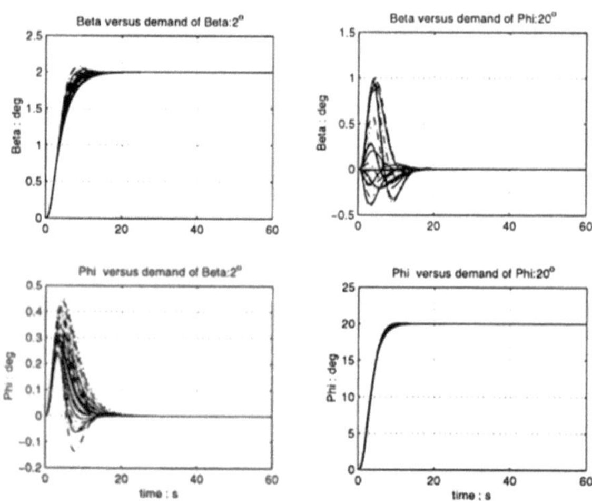

Figure 2.9. Proportional multi-model feedback, step responses of all RCAM models.

2.4.2 Multi-model design with a dynamic gain

The previous feedback design will be improved by considering a dynamic feedback. We shall follow the design procedure proposed in §2.1.4. More precisely, we propose the following steps:

- *Initial design.* We know from the previous analysis that the most important parameter is the mass. So, instead of considering the nominal model for the initial design, we shall consider a model corresponding to high mass (assuming that worst cases will correspond to low mass and, by the way, the risk of redundancy of multi-model constraints will be minimized, see page 114).

- *Analysis.* The previous control loop being closed, robustness analysis is performed (pole map or step responses). Worst case models are identified.

- *Multi-model design.* Improvement of the behavior of the worst case models keeping unaltered the properties obtained at the previous steps. This is a multi-model feedback design step.

- *Analysis.* New "worst case" model(s), and so on.

Initial feedback design. We will consider Model 12 for the initial design. We proceed as explained earlier (six dominant poles):

```
sys1 = rcamdata('lat',12,2,0.1);
pol  = [-0.9+0.9i; -0.8; -2.0; -1.5;-1.0 ];
KO   = fb_prop(sys1,0,pol,'n',[4 4 1 1 1])
```

Analysis of results. Using **rcampole** and **rcamstep**, the worst case is identified as being Model 25. Two pairs of complex poles of Model 25 controlled by KO are not well damped $\{-1.9845 + 4.4247i, -0.9887 + 3.0196i\}$. All other poles are well assigned.

Multi-model feedback design. The pairs of badly damped poles of Model 25 will be moved to the left with eigenvector projection in order to insure continuity. Before assignment, the gain structure is defined by invoking **str_cstr**, each coefficient has the following form

$$* + \frac{*s + *}{(s + 5 + 3j)(s + 5 - 3j)}$$

("*" denotes free parameters).

```
CSTR = str_cstr(2,6,-5+3i,0);
```

The criterion is defined by invoking **ktf_crit**. It consists of minimizing the distance of the gain being computed from KO.

```
CRIT = ktf_crit(CSTR,KO,i*[0 0.01 0.05 0.1 0.5]);
```

Eigenstructure assignment relative to `sys1`: all assignments are preserved (`new_pol1 = pol`).

```
pol1 = pol;
new_pol1=pol;
CSTR = eig_cstr(CSTR,sys1,new_pol1,'p',pol1,K0);
```

Eigenstructure assignment relative to `sys2`, the poles $\{-1.9845+4.4247i,\ -0.9887+3.0196i\}$ are shifted to $\{-4+4i, -3+3i\}$ (with eigenvector projection)

```
sys2 = rcamdata('lat',25,2,0.1);
pol2     =[ -1.9845+4.4247i ; -0.9887+3.0196i];
new_pol2 =[ -4+4i                ; -3+3i                ];
CSTR = eig_cstr(CSTR,sys2,new_pol2,'p',pol2,K0);
```

The gain is computed:

```
fb = fb_dyn(CSTR,CRIT);
```

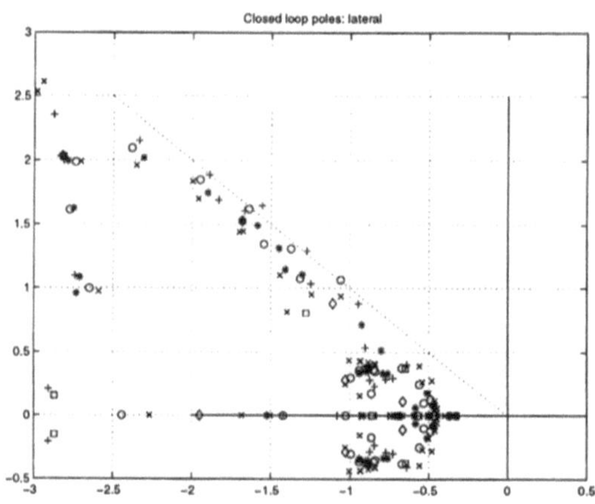

Figure 2.10. Dynamic multi-model feedback, poles of all RCAM models.

Final analysis:

```
rcampole('lat',fb,2,0);
axis([-3 0.5 -0.5 3])
```

Figure 2.10 shows that the results are perfect all over the flight envelope. We recognize below `new_pol1` in the first column (`eig_fb(sys1,fb)`) and `new_pol2` in the second one (`eig_fb(sys2,fb)`). In each column two pairs of complex conjugate poles more or less close to $-5+3i$ correspond to the denominators of the gain.

```
             -30.8280 +18.4427i  -31.8748 +19.9251i
             -30.8280 -18.4427i  -31.8748 -19.9251i
             -30.4916 +18.0204i  -31.1161 +18.9918i
             -30.4916 -18.0204i  -31.1161 -18.9918i
              -4.9135 + 3.6687i   -4.0000 + 4.0000i <- new_pol2
              -4.9135 - 3.6687i   -4.0000 - 4.0000i
              -4.7350 + 3.6947i   -2.6447 + 4.2773i
              -4.7350 - 3.6947i   -2.6447 - 4.2773i
new_pol1 ->   -0.9000 + 0.9000i   -2.3860 + 3.3820i
              -0.9000 - 0.9000i   -2.3860 - 3.3820i
new_pol1 ->   -2.0000             -3.0000 + 3.0000i <- new_pol2
new_pol1 ->   -1.5000             -3.0000 - 3.0000i
              -1.3501             -0.8496 + 0.3523i
new_pol1 ->   -0.8000             -0.8496 - 0.3523i
new_pol1 ->   -1.0000             -0.5706
              -0.9841             -0.3295
```

Comment. Improvements can be expected using structured feedback gains. For example, the degrees of the numerators should be less than the degrees of the denominators (roll-off not taken into account in the proposed solution).

2.4.3 Multi-model design using the tuning procedure

We use here the tuning procedure described in §1.4. The system considered is again the lateral model of the RCAM. After several design loops (not detailed here) following the design cycle described in §2.1.4.3, it appears that it is necessary to consider Models 0,3,11,15,12,10. These models are generated as follows:

```
sys1 = rcamdata('lat',0,2,0.07);
sys2 = rcamdata('lat',3,2,0.07);
sys3 = rcamdata('lat',11,2,0.07);
sys4 = rcamdata('lat',15,2,0.07);
sys5 = rcamdata('lat',12,2,0.07);
sys6 = rcamdata('lat',10,2,0.07);
```

Initial feedback design. We proceed as in §2.4.1 and 2.4.2.

```
pol = [-0.9+0.7i; -0.8; -2.0; -1.5; -1.0 ];
K0 = fb_prop(sys5,0,pol,'n',[4 4 1 1 1]);
```

The poles will be shifted into a trapezium the corners of which are the entries of the 2×2 matrix **Quad**. The parameter **Speed** must be tuned (large enough to accelerate tuning, small enough to prevent non-convergence) The parameter **Targ** is a point inside the quadrilateral.

```
Quad  = [-4+4*i -0.4+0.4*i;-4 -0.4];
Speed = 0.01;
Targ = -0.45 + 0.25*i;
```

Tuning of K0. For each considered models the eigenvalues assigned by the current value of K are computed (using **eig_fb**). Then, the poles with real part larger than -5 and non-negative imaginary part are selected using **sort_ev**. The function **cstr_qud** defines the first order constraints that shift the selected eigenvalues (if outside **Quad**) of an amount given by **Speed** towards[5] the point **Targ** (see Figure 1.15, page 85, for an illustration of the considered constraints).

```
K =K0;

for ii = 1:60;
   CSTR = cstr_ini(2,6);

   lam1 = eig_fb(sys1,K);
   lam1 = sort_ev(lam1,'r>-5&i>=0');
   CSTR = cstr_qud(CSTR,sys1,Quad,lam1,Speed,K,Targ);

   lam2 = eig_fb(sys2,K);
   lam2 = sort_ev(lam2,'r>-5&i>=0');
   CSTR = cstr_qud(CSTR,sys2,Quad,lam2,Speed,K,Targ);

   lam3 = eig_fb(sys3,K);
   lam3 = sort_ev(lam3,'r>-5&i>=0');
   CSTR = cstr_qud(CSTR,sys3,Quad,lam3,Speed,K,Targ);

   lam4 = eig_fb(sys4,K);
   lam4 = sort_ev(lam4,'r>-5&i>=0');
   CSTR = cstr_qud(CSTR,sys4,Quad,lam4,Speed,K,Targ);
```

```
   lam5 = eig_fb(sys5,K);
   lam5 = sort_ev(lam5,'r>-5&i>=0');
   CSTR = cstr_qud(CSTR,sys5,Quad,lam5,Speed,K,Targ);

   lam6 = eig_fb(sys6,K);
   lam6 = sort_ev(lam6,'r>-5&i>=0');
   CSTR = cstr_qud(CSTR,sys6,Quad,lam6,Speed,K,Targ);

   k = fb_tun(CSTR,[]);
   K = K + k;
end
```

The last two code lines update the gain K at each iteration. The empty input argument means that by default, it is the gain variation k that is minimized. The results are analyzed considering the pole map of Figure 2.11.

```
   rcampole('lat',K,2,0);
   axis([-3 0.5 -0.5 3])
```

There is not much improvement with respect to Figure 2.8, because there is not enough freedom in the proportional gain case. In fact, a dynamic gain must be used.

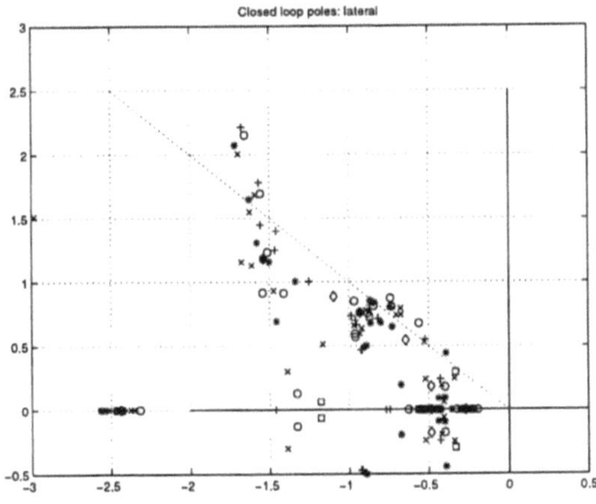

Figure 2.11. RCAM pole map corresponding to a proportional feedback obtained after tuning an initial law

2.5 Flexible systems control

Contents. The following possibilities for flexible structure control design will be addressed.

- Two pole assignment steps, one is considered at low frequencies, the other one at high frequencies ....§2.5.1

- Direct dynamic feedback ....§2.5.2

- MIMO generalized phase control ....§2.5.3

- Use of the tuning sub-toolbox with a dynamic gain .... §2.5.4

- Technique based on controllability analysis ....§2.5.5

References. Flexible control design in a multi-model setting is presented in [Le Gorrec et al., 1997]. The generalization of phase control to multivariable systems and to multi-model design is presented and illustrated in [Magni et al., 1997a].

Designing a control law for the considered system (the second one of Appendix §5.1) is not a difficult task because, adding in series a low pass filter (cut-off frequency $\omega_c \approx 6$ rd/s) before designing the rigid control law, suffices to prevent excitation of the flexibility. In order to illustrate several design possibilities, this filter will not be considered. The advantage of considering a so simple example is that each technique can be illustrated individually. For a much more complex control design problem (problem considered in [Chiappa et al., 2001a] which is multi-model, the system has about 100 states and very low frequency oscillatory modes) it was necessary to combine all the techniques presented here.

2.5.1   Low and high frequency feedback designs

A simple technique performed several times in distinct frequency bands leads sometimes to very good results. Here we will consider two bands.

1 - *Low frequency design.* Choose a frequency band $[0\ \omega_1]$ and add in series the corresponding low pass filter to the system. Then apply one of the design procedures of §1.2.

2 - *High frequency design.* Consider the closed-loop system obtained at the first step. Choose a frequency band $[\omega_2\ \omega_3]$, add in series the corresponding band pass filter. Then apply again one of the design procedures of §1.2. For high frequency control, the choice of eigenvectors by projection of open-loop ones is very efficient.

For the considered example the following choice will be made: $\omega_1 = \omega_3 = \infty$ and $\omega_2 = 10$ rd/s. The open-loop eigenvalues are

- rigid part: $\{-0.12 \pm 1.07i, -0.92, +0.0057, -10, -10\}$

- flexible part: $\{-0.31 \pm 11.5i, -0.34 \pm 17.7i, -1.66 \pm 28.1i\}$.

First, the gain Kr relative to the rigid part is computed:

```
sysr = demodata(1);
l1=-1;
l2=-1.2;
l3=-1.35+0.65*i;
Kr = fb_prop(sysr,0,[l1 l2 l3],'z',[1 1 4]);
```

When this feedback is applied to the flexible model (`sysf=demodata(2)`), the system becomes unstable. The eigenvalues are

- rigid part: $\{-1.28 \pm 0.65i, -1.30, -0.98, -9.16 \pm 0.94i\}$

- flexible part: $\{-0.30 \pm 11.1i, +0.29 \pm 16.9i, -1.31 \pm 27.3i\}$.

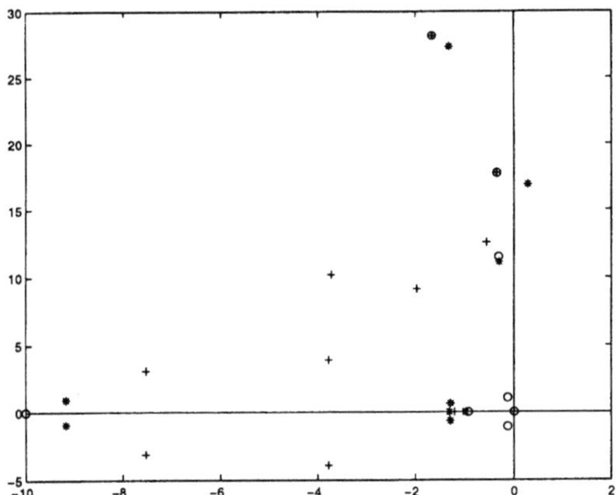

Figure 2.12. Poles: open-loop "o", rigid control "*", rigid plus flexible control "+".

A high pass filter with cut-off frequency ω_2 is added in series to each input of the system controlled by Kr.

```
sysf = demodata(2);

omegc = 10;
[ab,bb,cb,db] = butter(4,omegc,'high','s');
filt = ss(ab,bb,cb,db);

sysf1 = feedback(sysf,Kr,1)*[filt 0;0 filt];
```

Then, the two oscillatory poles $0.28+17j$ and $-1.3+27j$ are shifted to the left, the corresponding eigenvectors are chosen by eigenvector projection.

```
pol    = [-0.35+j*17.74,-1.66+j*28.14];
def_pb = [ 0.28+j*17   ,   -1.3+j*27   ];
Kf = fb_prop(sysf1,0,pol,'pp',def_pb)
```

The final gain is Kr + [filt 0;0 filt]*Kf. The eigenvalues are

- rigid part: $\{-1.29 \pm 0.64i, -1.20, -1.02, -14.4, -19.7\}$

- flexible part: $\{-0.55 \pm 12.6i, -0.35 \pm 17.7i, -1.66 \pm 28.1i\}$

- filters: $\{-7.53 \pm 3.11i, -3.73 \pm 10.22i, -3.78 \pm 3.92i, -1.97 \pm 9.17i\}$.

The results are summarized in Figure 2.12. The second feedback loop
has shifted back to their open-loop locations both destabilized oscillatory
poles.

2.5.2   Single step dynamic feedback design

The function **fb_dyn** was developed for multi-model design, but it can
also be used for single-model design problems as shown now. A dynamic
gain that assigns simultaneously rigid and flexible poles is designed in a
single step.

```
sysf = demodata(2);
```

The feedback gain is structured so that each of its entries have the following form

$$\frac{*s^2 + *s + *}{(s + 10 + 10j)(s + 10 - 10j)(s + 8)}$$

("$*$" denotes free parameters).

```
CSTR = str_cstr(2,4,[-10+10*i -8],1);
```

The criterion consists of minimizing the norm of the gain being computed, at some frequencies between 0 and 10 Rd/s.

```
CRIT = ktf_crit(CSTR,0,i*[0,0.01,0.1,1,10]);
```

The "rigid modes" will be assigned such that the 1st or 4th entry of the
corresponding right eigenvector are set to zero (option 'z'). The flexible
modes will be assigned by considering open-loop eigenvector projections
(option 'p'). "Rigid poles" to be assigned:

```
l1=-1;
l2=-1.2;
l3=-1.35+0.65*i;
```

open-loop "flexible poles" to be shifted

```
lf1 = -0.31+11.5i;
lf2 = -0.34+17.7i;
lf3 = -1.66+28.1i;
```

constraint definition using both options 'z' and 'p'

```
pol = [l1 l2 l3 lf1-0.5 lf2-0.5 lf3-0.1];
```

```
dpb = [1 1 4 1f1 1f2 1f3];
CSTR = eig_cstr(CSTR,sysf,pol,'zzzppp',dpb);
```

Finally, the feedback is computed by invoking fb_dyn.

```
fb = fb_dyn(CSTR,CRIT);
```

In the list of closed-loop poles given below (eig_fb(sysf,fb)), the 10 assigned poles are recognized.

```
 -1.7600 +28.1000i <- 1f3 - 0.2
 -1.7600 -28.1000i
 -0.8400 +17.7000i <- 1f2 - 0.5
 -0.8400 -17.7000i
-17.6634
-11.0427 +10.5794i
-11.0427 -10.5794i
 -8.0560 +12.0527i
 -8.0560 -12.0527i
 -0.8100 +11.5000i <- 1f1 - 0.5
 -0.8100 -11.5000i
-11.9668
 -1.1347 +  1.8135i <- 13
 -1.1347 -  1.8135i
 -1.3500 +  0.6500i
 -1.3500 -  0.6500i
 -1.0000            <- 11
 -1.2000            <- 12
```

2.5.3   Generalized phase control

Here, is illustrated the technique of §1.3.5 (page 70) applied to a single model (a similar but multi-model example is treated on page 74). The matrix Kr computed in §2.5.1 is considered as a preliminary feedback gain. The function dp_cstr permits us to define constraints for shifting back to the left the oscillatory poles. After the dynamic feedback gain fb is computed using fb_dyn, it will remain to tune the scalar loop gain rho (feedback gain equal to Kr + rho*fb).

The considered system is sysf1 defined in §2.5.1. The entries of the dynamic feedback have the following form

$$* + \frac{*s + *}{(s + 20 + 20j)(s - 20 + 20j)}$$

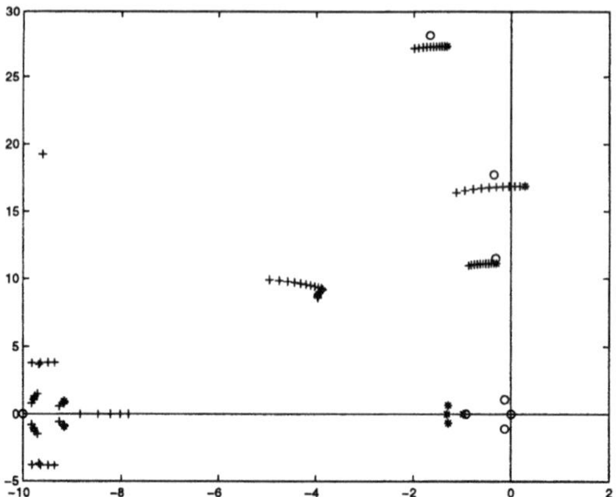

Figure 2.13.   Poles: open-loop "o", rigid control "∗", root locus "+".

("∗" denotes free parameters). The corresponding gain structure is defined by invoking **str_cstr**.

```
CSTR = str_cstr(2,4,-20+20*j);
```

The criterion is defined as

```
CRIT = ktf_crit(CSTR,0,i*[0 0.1 1 10]);
```

Constraints for "phase control": The flexible poles are shifted to the left on horizontal lines and the real and imaginary parts of $-1.3 + 0.65 * i$ ("rigid pole") are fixed. The arguments of **dp_cstr** defined below should be read column by column. For example, the first column means that the pole -1.31+27i must have its real ('r') part shifted to the left ('<') of at least -0.05.

```
lam =[-1.31+27i;0.28+16i;-0.30+11i;-9.1+0.93i;-1.3+0.65i];
fld1=[    'r'   ;   'r'   ;   'r'   ;   'r'    ;    'r'    ];
fld2=[    '<'   ;   '<'   ;   '<'   ;   '='    ;    '='    ];
fld3=[  -0.05   ; -0.10   ; -0.05   ;   0      ;    0      ];

lam =[lam ;-1.31+27i;.28+16i;-.30+11i;-9.1+.93i;-1.3+.65i];
fld1=[fld1;   'i'   ;  'i'  ;  'i'   ;   'i'    ;   'i'    ];
fld2=[fld2;   '='   ;  '='  ;  '='   ;   '='    ;   '='    ];
fld3=[fld3;    0    ;   0   ;   0    ;    0      ;    0     ];
```

```
CSTR = dp_cstr(CSTR,sysf1,lam,fld1,fld2,fld3);
```

Computation of the dynamic gain

```
fb = fb_dyn(CSTR,CRIT);
```

The results are given in Figure 2.13. The feedback is written as a function of the loop gain rho

```
fbrho = Kr+rho*[filt 0;0 filt]*fb;
```

The extremities of the root locus correspond to a loop gain rho equal to 10. For this value, the flexible poles are $\{-0.86\pm11j, -1.1\pm16.4j, -1.99\pm 27.1j\}$ which shows the reliability of the first order approximation used in the phase control approach in the case of flexible systems control.

2.5.4   Iterative technique

Applying the feedback gain Kr of §2.5.1 shifts to the right some oscillatory poles. In this section, a dynamic feedback initialized by Kr is tuned until oscillatory poles recover open-loop damping.

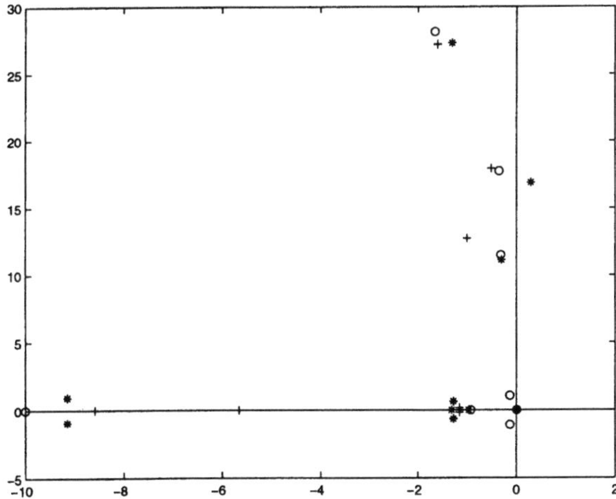

Figure 2.14.   Poles: open-loop "o", rigid control "*", after tuning gain coefficients "+".

First, the system and the initial gain are adapted to dynamic feedback tuning, see page 124 for explanations. The considered dynamic extension is of order 2. The gain corresponding to Kr after dynamic extension is slightly perturbed using the function smalldfb so that the initialization is not pathological (see page 88).

```
sysf = demodata(2);
dyn = [-3 -4];
K = smalldfb(sysf,dyn,Kr)
sysfdyn = add_dyn(sysf,dyn);
```

After this initialization, the following iterative procedure is proposed. The pole at high frequencies is moved to the left until its real part becomes less than -1.6, the one at intermediate frequencies, until its real part becomes less than -0.5 and the one at low frequencies, until its real part becomes less than -1. "Rigid poles" must not move. The function sort_ev is used for selecting modes in the above frequency ranges.

```
for ii = 1:100;

    lam  = eig_fb(sysfdyn,K);

    lam1 = sort_ev(lam,'i>=20&r>-1.6');
    CSTR = cstr_dp([],sysfdyn,lam1,'r','<',-0.02,K);

    lam2 = sort_ev(lam,'i>=15&i<20&r>-0.5');
    CSTR = cstr_dp(CSTR,sysfdyn,lam2,'r','<',-0.05,K);

    lam3 = sort_ev(lam,'i>=8&i<15&r>-1');
    CSTR = cstr_dp(CSTR,sysfdyn,lam3,'r','<',-0.05,K);

    lam4 = sort_ev(lam,'i<5&i>=0&r>-2');
    CSTR = cstr_dp(CSTR,sysfdyn,lam4,'r','=',0,K);
    CSTR = cstr_dp(CSTR,sysfdyn,lam4,'i','=',0,K);

    if length([lam1;lam2;lam3]) == 0;
       disp('BREAK');
       break;
    end;

    dk = fb_tun(CSTR,[]);
    K  = K+dk;

end;
```

See Figure 2.14 for the results.

2.5.5   Technique based on controllability analysis

The gain `Kr` computed in §2.5.1 is again considered. After closure
of this feedback loop, the input/output controllability of the poles is
analyzed using `plot_con`:

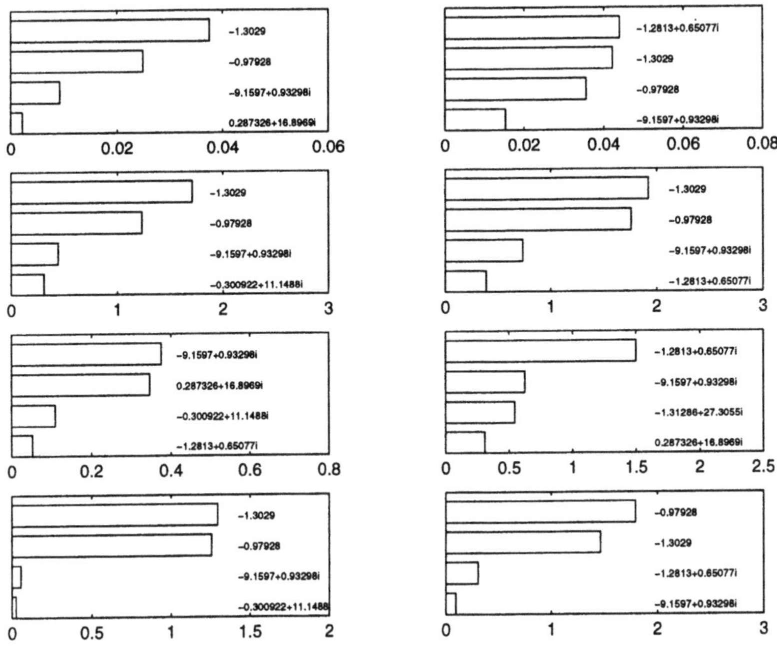

Figure 2.15.   Controllability analysis between all pairs of inputs - outputs.

```
sysffb=feedback(sysf,Kr,1);

figure
for ii = 1:4; for jj=1:2;
    subplot(4,2,(ii-1)*2+jj);
    plot_con(sysffb,jj,ii,4,6,'r-');
end;end;
```

The results are given in Figure 2.15. It turns out that the pole which
becomes unstable after closure of the "rigid loop" $(+0.29 \pm 16.9i)$ is well
controllable by the single gain between output number three and input
number one (see "`subplot(4,2,5)`"). Furthermore, the other pole which
has a similar degree of controllability $(-9.16 \pm 0.93i)$, is far from the
imaginary axis. So, it will suffice to tune this single gain to shift the

unstable mode with negligible effect on other poles. Reducing $Kr(1,3)$ from 6.58 to 3 leads to results very similar to those previously obtained.

2.6 Structured gain computation

Contents.

- Structured dynamic feedback ....§2.6.1
- Controller order reduction ....§2.6.2

References. The proposed multi-model design technique has shown to be very useful in practice for improving, structuring or simplifying existing controllers. Some problems more sophisticated than those treated in this manual can be found in [Le Gorrec et al., 1998a] (structured autopilot design), [Chiappa et al., 1998] (order reduction and robustness improvement of μ-synthesis-based design), [Chiappa et al., 2001a] (structuring and order reduction of a high dimensional $\mathcal{H}_2$ controller). Direct scheduled control design as in [Döll et al., 2000], can also be viewed as a multi-model structured feedback design technique.

The design problems of this section are purely academic. Realistic applications can be found in the aforementioned references.

2.6.1   Structured dynamic feedback

This section illustrates the computation of a feedback gain having a given structure. The lateral model of the RCAM will be considered:

```
sys = rcamdata('lat',0,1);
```

The feedback gain we look for has the following (purely academic) form

$$G(s) = \left[\begin{array}{cc} \frac{*}{s+3} \quad * + \frac{*s+*}{(s-9-9i)(s-9+9i)} & \frac{*}{10+s} \quad \frac{*}{10+s} \\ 0 \qquad \frac{*s+*}{(s-5-5i)(s-5+5i)} & \frac{*s+*}{(s-5-5i)(s-5+5i)} \quad 0 \end{array} \right]$$

(where "$*$" denotes the design parameters). The gain structure is defined by invoking the function **str_cstr**. The argument **polden** is a cell defining the denominator roots (entry per entry), the argument **difdeg** defines the difference of degrees between numerators and denominators (entry per entry).

```
polden = { [-3] [-9+9i] [-10] [-10];[] [-5+5i] [-5+5i] []};
difdeg = [ 1 0 1 1;1 1 1 1];
CSTR = str_cstr(2,4,polden,difdeg);
```

At this stage, it can be checked that the structure is well defined by invoking the function **str_view**. For example, type **str_view(CSTR,[1 1;1 2;1 3;1 4])**;.
In addition to gain structuring, it is required that $G_{(1,2)}(8j) = 0$.

```
CSTR = add_cstr(CSTR,[1 2],8,'=',0);
```

It remains to define a criterion (using **ktf_crit**),

```
CRIT = ktf_crit(CSTR,0,i*[0,0.01,0.05,0.1,0.5,1,5,10]);
```

to define the eigenstructure assignment constraints (same kind of constraints as in §2.3.1),

```
CSTR = eig_cstr(CSTR,sys,[-0.6+0.3i -0.6 -1.3],'n',[1 4 4]);
```

and to compute the feedback gain

```
[fb,X1,X2,ksi] = fb_dyn(CSTR,CRIT);
```

Note that the function **fb_view** can be used to display the results in transfer function form. For example, type **fb_view(CSTR,ksi,[1 1;1 2;1 3;1 4])**;. Analysis of the results:

```
figure
bode([1 0]*fb*[0;1;0;0])
```

The Bode locus showing that the frequency domain constraint is satisfied is given in Figure 2.16.

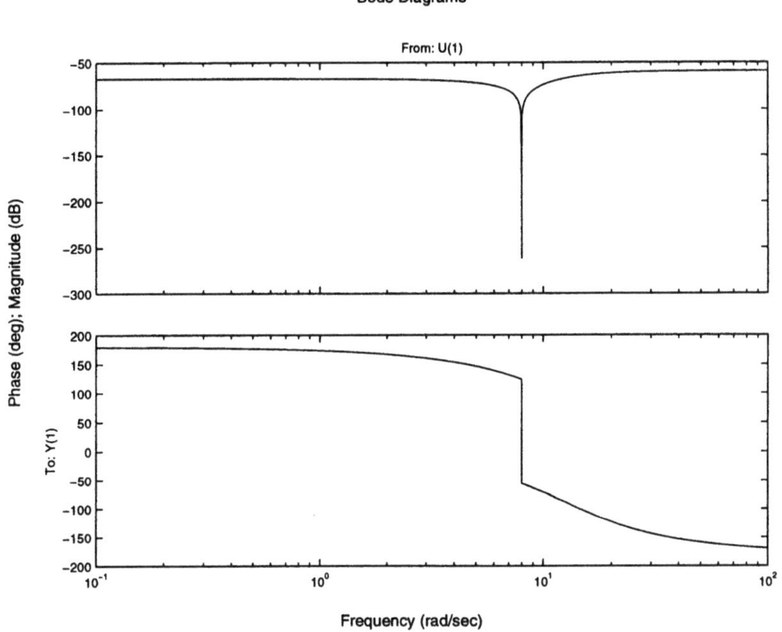

Figure 2.16. Bode locus of $G_{(1,2)}(j\omega)$.

In the list of closed-loop poles given below (**eig_fb(sys,fb)**), the 4 assigned poles are recognized. The other poles can be interpreted as a combination of non-dominant and controller poles.

```
 -8.9856 + 8.9906i
 -8.9856 - 8.9906i
-10.5272
 -6.0859 + 4.4114i
 -6.0859 - 4.4114i
```

```
-3.9065 + 2.7148i
-3.9065 - 2.7148i
-0.6000 + 0.3000i <- assigned
-0.6000 - 0.3000i
-1.3742
-1.3000                 <- assigned
-0.6000                 <- assigned
```

2.6.2   Controller order reduction

This section illustrates the controller order reduction technique described in §1.3.6 (page 72). A purely academic LQG feedback design relative to the RCAM is considered for order reduction.

```
sys = rcamdata('lat',0,1);
```

LQG design:

```
W = diag([1 1 1 1 1 1 0.2 0.2]);
V = diag([1 1 1 1 1 1 0.2 0.2 0.2 0.2]);
[a,b,c,d]=ssdata(sys);
[ac,bc,cc,dc] = lqg(a,b,c,d,W,V);
fb = - ss(ac,bc,cc,dc);
```

Feedforward design:

```
H = ff_stat(sys,fb,[1 4]);
```

Dominant closed-loop pole. This controller has 6 states. Step responses are considered in order to identify the dominant poles. Cross-coupling is ignored.

```
pol1 = lsim_mod(sysfb*H,'u1',60,0,'y1',2);
pol2 = lsim_mod(sysfb*H,'u2',60,0,'y4',2);
pol  = clean_ev([pol1;pol2],[],'r');
```

pol is the resulting vector of dominant poles (note that we have checked that there are not modes that cancel each other, see page 97).

Dominant feedback gain poles. Then, the dominant poles of the first and second row of fb are analyzed.

```
figure; plot_con(fb,[1:4],1);
figure; plot_con(fb,[1:4],2);
```

The first row of fb has a dominant pair of poles at $-2.6 + 0.05i$ (with weaker controllability than the poles of the second row). The second row has the following dominant poles $-5.3 + 4.4i$ and -7.2. In fact, the choice of denominators is not essential, this analysis is roughly taken into account. Two cases will be treated

- Case 1: $\mathbf{fb1} = \begin{bmatrix} *+\frac{*}{s+2} & \cdots & *+\frac{*}{s+2} \\ *+\frac{*}{s+7} & \cdots & *+\frac{*}{s+7} \end{bmatrix}$

- Case 2: $\mathbf{fb2} = \begin{bmatrix} * & \cdots & * \\ *+\frac{*}{s+7} & \cdots & *+\frac{*}{s+7} \end{bmatrix}$

(where "$*$" denotes the design parameters)

First case. The feedback structure is first defined by invoking str_cstr:

```
CSTR = str_cstr(2,4,{[-2];[-7]},0);
```

A criterion is defined by kft_crit in order to minimize the distance between fb and the feedback gain being computed (fb1) on a range of frequencies between 0 and 50 Rd/s:

```
CRIT = ktf_crit(CSTR,fb,i*[0,0.01,0.05,0.1,0.5,1,5,10,50]);
```

Finally, eigenstructure assignment constraints are defined in order to reassign the dominant eigenstructure that was originally assigned by fb:

```
CSTR = eig_cstr(CSTR,sys,pol,'p',pol,fb);
```

It remains to compute the corresponding feedback fb1:

```
fb1 = fb_dyn(CSTR,CRIT);
```

The feedback gain fb1 has 2 states (instead of 6 for fb). A comparison of the step responses is given Figure 2.17. It appears that the step responses considered for dominant pole analysis are perfectly reproduced. Cross-coupling responses are different, but this result is logical because this property was ignored in the dominance analysis. We did not take into account roll-off properties, in addition, the computed feedback gain is not strictly proper, therefore, the frequency domain properties cannot be recovered (more denominator poles are required for preserving roll-off properties).

Second case. We proceed as in the previous case:

```
CSTR = str_cstr(2,4,{[];[-7]},0);
CRIT = ktf_crit(CSTR,fb,i*[0,0.01,0.05,0.1,0.5,1,5,10,50]);
```

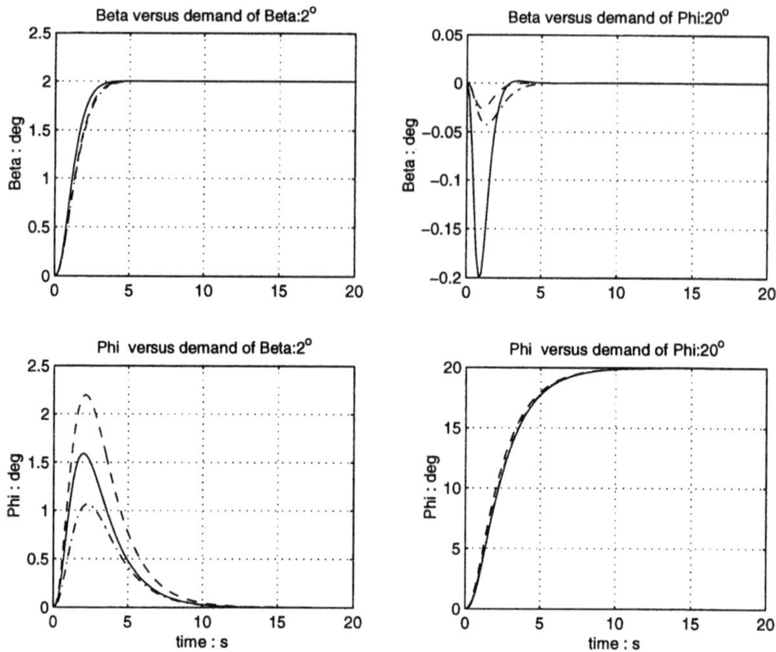

Figure 2.17. Comparison of LQG, and reduced order controller. (Solid lines: 6th order LQG controller, dashed lines: 2nd order controller, dash-dotted line: 1st order controller.

```
CSTR = eig_cstr(CSTR,sys,pol,'p',pol,fb);
fb2 = fb_dyn(CSTR,CRIT);
```

The feedback gain `fb2` has 1 state (instead of 6 in `fb`). Conclusions are similar as above.

Comment. It is not easy to define a systematic controller order reduction procedure because the proposed technique consists of a re-design which depends on the original design specifications. More precisely, the selection of "dominant poles" depends on the properties that must be recovered after order reduction. Furthermore, when some pairs of modes appearing as highly dominant cancel each other in the modal decomposition, an expertise of the designer is necessary to decide of ignoring them (see page 97). If the criterion is not sufficient to recover frequency domain properties, analysis of the "dominant poles" in the frequency response must also be considered.

Notes

1 This effect can be explained by the fact that, projection with small change of eigenvalue, is equivalent to the first order pole motion of Equation (1.99).

2 This technique is somewhat iterative, but much less than the other one as three or four iterations are often sufficient, and at each step the involved computation is systematic (Linear Quadratic Programming).

3 The wording "damping ratio" is only meaningful in the SISO case with numerators equal to the unity. But, it is also currently misused in the MIMO case, just for characterizing pole location.

4 Pole assignment with multiplicity higher than one comes from open-loop poles having the same imaginary part (real poles for example). Repeated poles may lead to numerical problems when some other functions of this toolbox are invoked.

5 The `Targ` point is denoted "T" in Figure 1.15. Depending on the location of this point, for example close to the corner Q11 or close to the corner Q12, the convergence properties might be very different.

Chapter 3

TOOLBOX REFERENCE

Contents.

- List of available functions ....page 169
- Help messages ....page 173

3.1 List of available functions

Core function	
defin_vw	Closed-loop eigenstructure selection

Proportional feedback design	
fb_prop	General eigenvalue / eigenvector assignment
sfb_ins	Insensitive state feedback
sfb_proj	State feedback assignment by projection

Observer manipulation and design	
ob_gene	General observer design by standard assignment
ob_ins	Insensitive observer design
sob_proj	Full state observer design by projection
add_obs	Connects an observer
obs2dfb	Observer-based fb → standard dynamic fb
dfb2obs	Standard dynamic fb → observer-based fb

Single-model dynamic feedback design	
fb_dyn	Special case of multi-model case, see next page
dfb_ins	Observer-based insensitive dynamic feedback
dfb_proj	Observer-based dynamic feedback using projections

Dynamic and proportional feedback (to/from)	
add_dyn	To treat dynamic feedback with fb_prop or fb_tun
sta2dyn	Gain feedback → dynamic feedback (see add_dyn)
dyn2sta	Dynamic feedback → gain feedback (see add_dyn)

Feedforward design	
ff_assgn	"Assignment" by feedforward/cancellation
ff_stat	Static gain made equal to identity

Multi-model dynamic feedback	
Multi-model sub-toolbox	
`fb_dyn`	Main function (treatment of CRIT and CSTR)
`fb_view`	Displays results to the screen
Building of the matrix CRIT	
`ktf_crit`	Definition of a criterion relative to the gain
Building of the matrix CSTR	
`str_cstr`	Definition of gain structure
`str_view`	Displays gain structure to the screen
`eig_cstr`	Multi-model eigenstructure assignment
`dp_cstr`	Pole shifting, multi-model case
`add_cstr`	Frequency domain template for gain shaping

Tuning of proportional feedback	
Tuning sub-toolbox	
`fb_tun`	Main function (treatment of CRIT and CSTR)
Building of the matrix CRIT	
`crit_k`	Crit. relative to the entries of $K_0 + \Delta K$ or ΔK
`crit_ctr`	Equality constraints replaced by a criterion
Building of the matrix CSTR	
`cstr_ini`	Initialization of the matrix CSTR
`cstr_k`	Constraints on gain coefficients
`cstr_dp`	Constraints for pole shifting
`cstr_ev`	Constraints for eigenvalue assignment
`cstr_eig`	Constraints for eigenstucture assignment
`cstr_qud`	Constraints for shifting poles into quadrilaterals
`dist_qud`	Distance of poles from a quadrilateral
Building of the matrix CSTR by the designer	
`comp_dv`	First order eigenstructure (v, w, u, t) variation
`k2ksi`	Treatment of the trace of a product of matrices
`ab2cstr`	Plugs constraints into the matrix CSTR

Selection of eigenvalues	
`choi_ev`	Ordering of eigenvalues
`clean_ev`	Treats complex conjugates
`sort_ev`	Extracts eigenvalues satisfying a logical expression
`contr_ev`	Selects eigenvalues with minimum controllability
`vicin_ev`	Extracts eigenvalues in the vicinity of given values

Functions for modal analysis	
`lsim_mod`	Modal simulation and plotting
`plot_con`	Controllability / observability degrees measure
`plot_res`	Residuals
`azer`	Zero and almost-zero computation
`plot_zer`	3D-plot for almost-zeros
`eig_fb`	Closed-loop eigenstructure computation

Demonstration	
`rmctdemo`	Opens a window for demonstration
`demodata`	Miscellaneous models
`rcamdata`	Bank of aircraft linearized models
`rcampole`	Pole analysis relative to `rcamdata`
`rcamstep`	Step response analysis relative to `rcamdata`

3.2   Help messages

Comments. The optional arguments of each function are given between square brackets at the beginning of each **help1** message. Example:

```
[[x,]y] = foo(a[,b,c[,d]]);
```

Corresponding correct command lines:

```
y = foo(a);
y = foo(a,b,c);
y = foo(a,b,c,d);
[x,y] = foo(a);
[x,y] = foo(a,b,c);
[x,y] = foo(a,b,c,d);
```

Remember to use **help1** and **help2** instead of **help**.

3.2.1 Function: **ADD_CSTR**

Purpose. Defines constraints relative to frequency domain shaping of the feedback gain in a form that can be treated by **fb_dyn**.

Synopsis.

```
CSTR = add_cstr(CSTR,IJ,fld1,fld2,fld3[,fb0]);
```

Description. First, read page 208. The constraints dealt with are:

- modulus of the feedback gain at 0 rd/s, larger than or equal to a given positive value

- modulus of the feedback gain at a specified frequency, less than a given positive value

- modulus of the feedback gain at a specified frequency, equal to zero.

Other constraints relative to the vicinity with a reference feedback can be considered. This function belongs to the multi-model sub-toolbox. The matrix CSTR must be initialized (see **str_cstr**) before invoking this function.

Input arguments.

CSTR   Set of existing constraints when **add_cstr** is called.

IJ   Matrix having two columns defining the entries of the gain coefficients that are treated (see example below). IJ,fld1,fld2,fld3 have the same number of rows.

fld1   Vector of real numbers specifying the frequencies (rd/s) at which constrtaints are considered.

fld2   String vector, each entry of fld2 is '=', '<' or '>'.

fld3   Vector of real numbers specifying the constraints objectives. For example:

```
IJ = [1 1 ; 1 2];
fld1 = [ 0 ; 10 ];
fld2 = [ '>'; '='];
fld3 = [ 15 ; 0 ];
```

means that the modulus of K(1,1) > 15 at 0 rd/s and that the modulus of K(1,2) = 0 at 10 rd/s.

fb0   Reference feedback for constraints definition (constant matrix or ss-object). If `fb0` is given as input argument `fld3` gives the percentages of variation w.r.t. to this reference gain. Example:

```
IJ = 'all';
fld1 = 0;
fld2 = '<';
fld3 = 10;
```

means that at 0 rd/s, all the entries of the computed feedback gain should not vary of more than 10 percent w.r.t. the reference feedback `fb0`.

Output argument.

CSTR   Input constraints plus new constraints computed by `add_cstr`. To be used as input argument of `fb_dyn`, `eig_cstr` or `dp_cstr`

See also: `fb_dyn`, `eig_cstr`, `dp_cstr`, `str_cstr`

Examples. Illustrative examples are given on page 76 and page 160. The example of page 76 proposes the definition of frequency domain feedback gain constraints together with multi-model "phase control" constraints. In the other example, an entry of the feedback gain is constrained to zero at a given frequency.

Discussion. These constraints are defined on page 68, more details can be found in [Magni, 1999]. Constraints are considered independently for each entry of the gain. In addition inequality constraints relative to the norm are replaced by the four (linear) inequalities of (1.89), page 69. However, it is possible as shown in [Magni et al., 1998], to consider the standard norm (at fixed frequencies) but in this case the problem for finding the gain becomes an LMI instead of a LQP problem. (The structure of the resulting matrix CSTR is given on page 295).

3.2.2   Function: ADD_DYN

Purpose. Adds integrators in parallel to a given system so that a dynamic feedback relative to the original system can be viewed as a proportional feedback relative to the extended system.

Synopsis.
```
sys1 = add_dyn(sys0,dyn);
[a1,b1,c1,d1] = add_dyn(a0,b0,c0,d0,dyn);
```

Description. If the original system `sys0` corresponds to the state-space representation (A_0, B_0, C_0, D_0), the augmented system `sys1` is given by

$$\left(\begin{bmatrix} A_0 & 0 \\ 0 & \text{Diag(dyn)} \end{bmatrix} ; \begin{bmatrix} B_0 & 0 \\ 0 & I \end{bmatrix} ; \begin{bmatrix} C_0 & 0 \\ 0 & I \end{bmatrix} ; \begin{bmatrix} D_0 & 0 \\ 0 & 0 \end{bmatrix} \right)$$

(assuming that `dyn` is a real vector, otherwise the diagonal matrix is replaced by a Shur block-diagonal one). This function must be used with `dyn2sta` and `sta2dyn`. This set of three functions permits the designer to tune dynamic feedback gains using `fb_tun` (which treats proportional feedback gains) or to design dynamic controllers using `fb_prop` (devoted to proportional gain design).

Input arguments.

a0,b0,c0,d0 **or** sys0   Initial system.

dyn   Vector of additional dynamics without repetition of complex conjugate values.

Output arguments.

a1,b1,c1,d1 **or** sys1   Extended equivalent system.

See also: `sta2dyn, dyn2sta`

Examples. See pages 46, 145 and 205. The example of page 46 illustrates the use of a dynamic extension for exact pole assignment by output feedback. The one of page 145 illustrates dynamic feedback gain tuning using `fb_tun`.

Discussion. See page 45. Note that usually, the additional dynamic states correspond to pure integrators. Here, the input argument `dyn` permits the designer to consider first or second order transfer functions instead of integrators.

3.2.3   Function: ADD_OBS

Purpose. Connects an observer to a system.

Synopsis.
```
[sysobs,sysu] = add_obs(sys,U,T,Q,Pi);
[sysobs,sysu] = add_obs(sys[,sysnom],U,T,Q[,Pi]);
```

Description. Connects the observer described by U, T, Q to the possibly non-nominal system **sys**. The observer is relative to the nominal system **sysnom**. See the figure below for the definition of U, T, Q. Note that the matrix Pi of this figure is computed by **add_obs** from U and T.

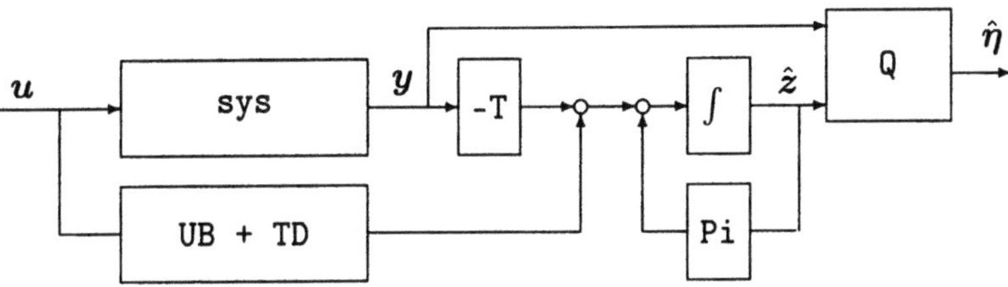

Input arguments.

sys   LTI system (see **ss.m**).

sysnom   LTI system. It is the system which was used for designing the observer (*i.e.*, for computing U and T). By default **sysnom** = **sys**.

U,T   Matrices characterizing observers.

Q   Matrix mixing measurements and observer states:

- State observer with Kalman filter structure: U is n by n (n is the number of states) and Q = [zeros(size(T)) inv(U)]

- Minimum order state observer: [C;U] is n by n and Q = inv([C;U])

- To observe L x consider Q s.t. Q*[C;U] = L

- To observe [C;U] x consider Q as the identity matrix (short cut: Q = 1).

Pi   This input argument is useful only if the observer order is larger than the number of states because in this case Pi is not unique.

Output arguments.

sysobs   System **sys** with connected observer. The input of the connected systems is the initial input. The output is Q[y;z] where y is the initial measurement vector and z the vector of the observer states (*i.e.*, estimate of Ux).

sysu   System **sys** but with Q[y;Ux] as measurement vector.

See also: `ob_ins`, `ob_gene`, `sob_proj`

Examples. See pages 58 (second design), 135, 137 and 219. Pages 58 and 135 is illustrated the fact that we can first connect the observer to the system and then, design an output feedback for the composite system.

Discussion. See pages 47, 54 and Figure 1.7 (Q is the matrix $[Q_y \; Q_z]$ of this figure).

3.2.4   Function: AZER

Purpose. Computes the almost invariant zeros of a system.

Synopsis.

```
[alzer,jsvd] = azer(sys[,rad[,tol[,zer0]]]);
```

Description. Almost zeros are local minima of the minimum singular value of the system matrix relative to `sys`. To decide if the values in `alzer` are true zeros (minimum equal to zero), relevant almost zeros or irrelevant "almost zeros", the user must analyze the values of the local minima (given in `jsvd`). If the values of almost zeros are far from dominant poles, these values must be ignored (like zeros at infinity). Note that the results might vary from on run to another on account of the fact that the used algorithm is initialized by random high gains.

Input arguments.

sys   LTI system (see `ss.m`).

rad   Radius of a circle around origin in which almost zeros are considered. Default `Rad = 100`.

tol   Tolerance to distinguish between almost zeros. Default, `tol = 1e-3`.

zer0   Vector of initial values for the search of almost zeros. Such initial values can be selected using `plot_zer`.

Output arguments.

alzer Almost zeros.

jsvd   Corresponding minimum singular value of system matrix. `jsvd(i)` large $\Rightarrow$ `alzer(i)` cannot be considered as a zero.

See also: `plot_zer`

Example and references. See page 102. See also `plot_zer` for validation of almost zeros by 3-D plotting.

3.2.5   Function: CHOI_EV

Purpose. Recognizes the entries of a vector within the entries of another given one. The entries to be recognized in both vectors do not need to be equal but must to be close enough.

Synopsis.

`[lamo,Vo,Wo,Uo,To] = choi_ev(lam0,lami[,Vi[,Wi[,Ui[,Ti]]]])`

Description. If the second input argument (`lami`) is a list of eigenvalues, it is possible to consider as additional input arguments: the right eigenvectors (`Vi`), the input directions (`Wi`), the left eigenvectors (`Ui`) and the output directions (`Ti`) (see `eig_fb`). In this case, the eigenvalues of `lami` are selected and re-ordered following `lam0`. A similar selection and re-ordering is made for the columns of `Vi`, `Wi` and for the rows of `Ui`, `Ti`.

Input arguments.

`lam0`   Vectors of complex numbers to be recognized in `lami`.

`lam`   Vector of complex numbers (eigenvalues for example).

`Vi,Wi,Ui,Ti`   Eigenstructure corresponding to `lami`. Can be generated by calling `eig_fb`.

Output arguments.

`lamo`   Such that `lamo(i)` belongs to `lami` and is close to `lam0(i)`.

`Vo,Wo,Uo,To`   are columns or rows of `Vi,Wi,Ui,Ti` re-ordered like `lamo` with respect to `lami`.

See also: `sort_ev, vicin_ev, eig_fb, clean_ev`

Example. This function can be ignored as it is mostly called by other functions of the toolbox. However, in some cases it can be useful, for example, in order to follow pole motions when the gain is tuned. This point is illustrated now. In order to classify the eigenvalues of (A +

B*K0*C) and of (A + B*(K0 + k)*C) in the same order (where k is a small variation of K0).

```
randn('seed',0); rand('seed',0);
sys = rss(8,4,4);
K0 = ones(4,4); k = 0.1*rand(4,4);
lam0 = eig_fb(sys,K0);
```

```
[lam0 choi_ev(lam0,eig_fb(sys,K0+k))]
```

The result is

```
-5.4337                -5.3537
-3.4425                -3.4879
-3.2058                -3.2078
-0.3946 + 0.6314i      -0.1718 + 1.3395i
-0.3946 - 0.6314i      -0.1718 - 1.3395i
-1.4580                -1.5790
-1.0995                -1.1477
-0.9895                -1.0208
```

The second column gives the closed-loop eigenvalues after feedback perturbation (`eig_fb(sys,K0+k)`). The entries of this column are ordered so that it is possible to recognize in the same order the eigenvalues given in the first one. This function can also be used to recognize a part of the spectrum, for example `choi_ev(lam0(1),eig_fb(sys,K0+k))` identifies -5.3537 as being the perturbed value of -5.4337. However this function should be used with perturbations having a smaller effect on the closed-loop spectrum. In particular, it does not work when some eigenvalues change from real to nonreal or conversely.

If the eigenstructure is also of interest, use for example:

```
lam0 = eig_fb(sys,K0);
[lami,Vi,Wi,Ui,Ti] = eig_fb(sys,K0+k);
[lamo,Vo,Wo,Uo,To] = choi_ev(lam0,lami,Vi,Wi,Ui,Ti);
```

3.2.6   Function: CLEAN_EV

Purpose. Treats a vector of complex numbers in such a way that complex conjugate values are not repeated, or on the contrary, the value with negative imaginary part is given just beneath the one with positive imaginary part. Can also be used for removing duplication.

Synopsis.

```
[beta1,beta2] = clean_ev(beta[,tol[,option]])
```

Input arguments.

beta   Vector to be treated.

tol   Tolerance for equality of entries of beta.

option   String equal to 'k' for keeping duplication in beta (default) or equal to 'r' for removing duplication in beta.

Output arguments.

beta1   Vector beta without repeated conjugate values.

beta2   Vector beta with repeated conjugate values.

See also: sort_ev, vicin_ev, choi_ev, contr_ev

Discussion. This function is mostly called by other functions of the toolbox. However, it can be used for example as mentioned on page 73, in order to remove duplication in a vector of dominant modes obtained after several runs of lsim_mod, plot_res and / or plot_con (see § 2.6.2).

3.2.7   Function: COMP_DV

Purpose. Computes the variations of the right and left eigenvectors, input and output directions, that are induced by a feedback gain variation.

Synopsis.

[lam,v,w,u,t,Xv,Yv,Xw,Xu,Yu,Yt] = comp_dv(lambda,sys,K0);

Description. Let k, v, w, u, and t denote respectively the variation of K0, of the right and left eigenvectors, of the input and output directions. The variations of these vectors are given by:

    dv = Xv*k*Yv; (v'*v = 1)
    dw = Xw*k*Yv;
    du = Xu*k*Yu - u*Xv*k*Yv*u; (u*v = 1)
    dt = Xu*k*Yt - u*Xv*k*Yv*t;

The constraints generated using the output arguments of comp_dv can be added to those in CSTR by using k2ksi and ab2cstr (see also cstr_ini for initializing CSTR). This function belongs to the tuning sub-toolbox.

Input arguments.

lambda   Scalar number that is close enough to one of the entries of eig_fb(sys,K0) so that it can be recognized without ambiguity in this spectrum.

sys   LTI system (see `ss.m`).

K0   Initial proportional gain.

Output arguments.

lam   Is the (exact) closed-loop eigenvalue the closest to the input argument `lambda`.

v,w,u,t   Corresponding modal vectors.

Xv,Yv,Xw,Xu,Yu,Yt   See above.

See also: `k2ksi, ksi2k, ab2cstr, cstr_ini`

Examples. The example of page 92 illustrates an output feedback design (by tuning the feedback gain) with eigenvalue sensitivity minimization. Let us consider an example consisting of plugging into the matrix CSTR a constraint relative to the gain variation of the form $d(uXv) < -0.1$, where u and v are left and right eigenvectors and X is a given fixed matrix. First, we have to find the expression of $d(uXv) = du\ X\ v + u\ X\ dv$. Using the notations given in the "description" paragraph, the above variation is given by
Xu*k*Yu*X*v - u*Xv*k*Yv*u*X*v + u*X*Xv*k*Yv.
As it is a scalar, we can use the trace operator and permute matrices so that the above variation becomes trace((Yu*X*v*Xu - Yv*u*X*v*u*Xv + Yv*u*X*Xv)*k). Therefore, let us consider the following command lines:

```
X = rand(6,6); sys = demodata(1);

[lam,v,w,u,t,Xv,Yv,Xw,Xu,Yu,Yt] = comp_dv(-0.92,sys,0);
M = Yu*X*v*Xu - Yv*u*X*v*u*Xv + Yv*u*X*Xv;
```

It remains to plug the constraint into the matrix CSTR (which is initialized here using `cstr_ini`):

```
A = k2ksi(M')'; b = -0.1;
CSTR = cstr_ini(2,4);
CSTR = ab2cstr(CSTR,[],[],A,b);
```

Discussion. This function permits the designer to compute the gradient (with respect to gain variations) of all kinds of formulas in which appear the right and left eigenvectors or the input and output directions. Details are given in §1.4.4. See also [Magni and Manouan, 1994].

3.2.8   Function: CONTR_EV

Purpose. Sorts out the entries of a vector of closed-loop eigenvalues with degree of controllability larger than a given bound.

Synopsis.

```
ll = contr_ev(sys,K,lambda,bound);
```

Input arguments.

sys   LTI system (see `ss.m`).

K   Proportional feedback.

lambda   Vector of the closed-loop eigenvalues that are to be tested.

bound   Controllability bound for selecting entries of `lambda`. In order to choose a value for `bound` analyze the results of `plot_con`.

Output argument.

ll   Selected values of `lambda`.

See also: `clean_ev`, `choi_ev`, `sort_ev`

Discussion. This function should be used in order to prevent attempts to shift (using the tuning sub-toolbox) weakly controllable eigenvalues. See §1.5.2 page 99 and see more precisely the comments given at the end of this section. It is considered in `sfb_proj`, `obs_proj` and `dfb_proj` in order to prevent re-assignment of weakly controllable poles.

3.2.9   Function: CRIT_CTR

Purpose. Transforms some selected equality constraints into a quadratic criterion (see `fb_tun`).

Synopsis.

```
[CRIT,CSTR] = crit_ctr(CSTR,neqcrit,CRIT)
```

Description. First, read page 211. Let us denote `Ax=b` the `neqcrit` first equality constraints contained in `CSTR`. These constraints are replaced by the criterion $norm(Ax-b)^2$. Other constraints, if any, remain in `CSTR`. This new criterion is added to the one already contained in the input arguments `CRIT`. Transformation of constraints into a criterion is useful for avoiding overly stringent sets of equality constraints. This function belongs to the tuning sub-toolbox. It is not compatible with the multi-model sub-toolbox.

Input arguments.

CSTR   Output argument of other functions of the tuning sub-toolbox.

neqcrit   Number of concerned equality constraints of CSTR.

CRIT   Initial criterion.

Output arguments.

CSTR   New set of constraints without the ones moved to the criterion.

CRIT   Sum of the criterion in input argument and of the criterion replacing the transformed constraints.

See also: crit_k, fb_tun

Example. The following example considers an alternative version of the illustrative example relative to cstr_dp (see page 186). The inequality constraints are replaced by equalities and then, by a criterion added to the existing one.

```
sys = demodata(1);
k1 = zeros(2,4);
CRITO = crit_k(ones(2,4),zeros(2,4));

  for ii = 1:100;
  % Eigenvalues to be shifted
     lam = eig_fb(sys,k1);
     lamshift = sort_ev(lam,'r','>',-0.5,'&','i','>=',0);
     if length(lamshift) == 0; disp('BREAK'), break,end;
  % Shift (equality constraint)
     CSTR = cstr_dp([],sys,lamshift,'r','=',-0.1,k1);
  % Equality constraint replaced by a criterion
     neq = infocstr(CSTR);
     [CRIT,CSTR]  = crit_ctr(CSTR,neq,CRITO);
  % The gain is updated
     dk = fb_tun(CSTR,CRIT);
     k1 = k1+dk;
  end;

  [eig_fb(sys) eig_fb(sys,k1)]
```

The result is:

```
-0.1272 + 1.0734i  -9.3215
-0.1272 - 1.0734i  -0.5043 + 0.6242i
-0.9212            -0.5043 - 0.6242i
 0.0057            -0.5029 + 0.0667i
-9.9000            -0.5029 - 0.0667i
-10.1000           -9.8341
```

This example is only given for illustration: the function `crit_ctr` must be considered for more complex problems when the set of constraints becomes overly stringent.

3.2.10   Function: CRIT_K

Purpose. Defines a criterion to be minimized by the function `fb_tun`.

Synopsis.

```
CRIT = crit_k(Kweight,Kc)
```

Description. First, read page 211. Three cases are of interest.

- The gain variation k is to be minimized, then Kc = 0.

- The sum K0 + k is to be minimized (K0 initial gain), then the second input argument is K0.

- If it is the difference between the gain K0 + k and a given reference gain Kref that must be minimized, the second input argument should be K0 - Kref (corresponding criterion relative to K0 - Kref + k).

The considered criterion is

$$J = \frac{1}{2} \sum_{i,j} (\texttt{Kweight(i,j)} \ (\texttt{Kc(i,j)} + \texttt{k(i,j)}))^2$$

It is possible to combine two criteria as follows: CRIT = 11*CRIT1 + 12*CRIT2 where for example CRIT1 = `crit_k(Kweight,0*K0)` and CRIT2 = `crit_k(Kweight,K0)`. This function belongs to the tuning sub-toolbox.

Input arguments.

Kweight   Matrix with same size as Kc and all entries positive.

Kc   Matrix for defining the criterion (see above).

Output argument.

CRIT   Used for communication between RMCT functions.

See also: `crit_ctr`, `fb_tun`

Examples. See pages 89 and 92. In both cases, this function is used to prevent large variations of the gain within a tuning loop.

Discussion. Using a large non-zero second input argument might induce large variations that are not compatible with the first order approximations used for tuning, see page 81 for more details. The content of the output argument is described page 295.

3.2.11   Function: CSTR_DP

Purpose. Defines constraints for shifting eigenvalues (see `fb_tun`).

Synopsis.
```
CSTR = cstr_dp(CSTR,sys,lam,fld1,fld2,fld3,K0)
```

Description. First, read page 211. `cstr_dp` defines first order constraints for pole shifting. The considered constraints are relative to the real and imaginary part variations. The variations can also be constrained on a line with a given slope. This function belongs to the tuning sub-toolbox.

Input arguments.

CSTR   Matrix of existing constraints when `cstr_dp` is called.

sys   LTI system (see `ss.m`).

lam, fld1, fld2, fld3   These four vectors have the same length. They are used for defining pole shifting constraints. If all rows of `fld1`, `fld2`, `fld3` are equal, these arguments can be (1 by 1).

lam   Vector of the poles of `sys` controlled by the feedback K0.

fld1   Vector containing the following strings 'r', 'i' or 's' (for real part, imaginary part and slope).

fld2   Vector containing the following strings '<', '=' or '>'.

fld3   Vector containing numerical values.

K0   Initial proportional gain. Example:
```
lam = [lam1; lam1;lam2];
fld1 = ['r';'i';'s'];
fld2 = ['<';'=';'='];
fld3 = [-0.1; 0 ; 1.1];
```
means that the variations d(lam1), d(lam2) of the closed-loop eigen-
values lam1, lam2 are such that:
real part of d(lam1) less than -0.1
imaginary part of d(lam1) equal to 0
imaginary part of d(lam2) equal to 1.1 times its real part.

Output argument.

CSTR Used for communication between RMCT functions.

See also: `fb_tun`, `cstr_qud`, `cstr_eig`, `cstr_ev`

Example 1: Let us define first order constraints for shifting two poles
of $A + BK_0C$. The pole of $A + BK_0C$ the closest to $-1 + 2i$ is shifted to
the left of at least -0.15 with unchanged imaginary part. The pole the
closest to -0.2 is shifted to the left of an amount exactly equal to -0.1.
```
lam  = [-1+2i ; -1+2i ; -0.2];
CSTR = cstr_dp([],sys,lam,'rir','<==',[-0.15;0;-0.1],K0);
```

Example 2: Two eigenvalues are shifted to the 0.7 damping ratio area
along lines of slope $= + 0.4$. First is defined a system:
```
a = [-5 8 0 0 2;-8 -5 1 1 2;0 0 -3 4 0;0 0 -4 -3 0;...
        0 0 3 4 -5];
b = [1 0;0 1;1 0;0 1;1 0]; c=[eye(3,3) zeros(3,2)];
sys = ss(a,b,c,0);
k0 = zeros(2,3);
polplot = eig(a);
```
The proposed tuning loop is as follows:
```
for ii = 1:60;
    lam = eig_fb(sys,k0);
% Selection of eigenvalues to be shifted
    lamshift = sort_ev(lam,'d<0.7&i>=0');
    if length(lamshift) == 0; disp('BREAK'), break; end;
% Pole shifting
%   real part move at least of -0.05
    CSTR = cstr_dp([],sys,lamshift,'r','<',-0.05,k0);
%   motion along a line of slope = + 0.4
    CSTR = cstr_dp(CSTR,sys,lamshift,'s','=',0.4,k0);
```

```
% The gain is updated
    dk = fb_tun(CSTR,[]);
    k0 = k0+dk;
    polplot=[polplot;eig_fb(sys,k0)];
end;
```

Analysis of the results:

```
figure
plot(real(polplot),imag(polplot),'k+')
hold on
plot([0 -10],[0 10],'k-')
axis([-10 0 0 10])
```

The resulting pole motion is illustrated as follows

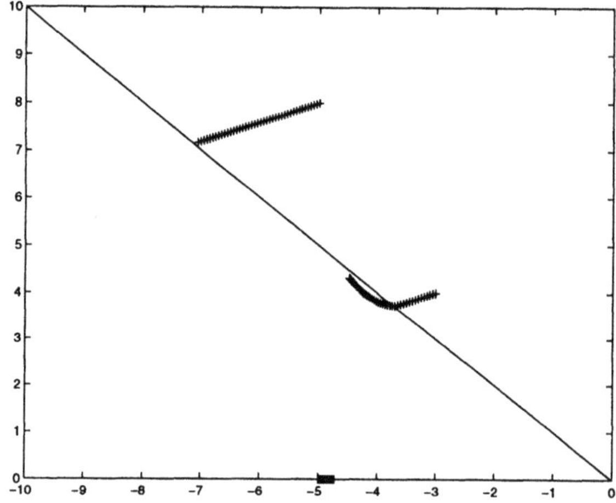

For an other example, see page 89.

Discussion. See page 82 (and [Magni and Manouan, 1994]). See also page 295 for the structure of the resulting matrix CSTR. Note that `cstr_dp` looks like `dp_cstr`, however the latter function is relative to a dynamic feedback; it cannot be used within a tuning loop as the order of the dynamic gain would increase at each step.

3.2.12   Function: CSTR_EIG

Purpose. Defines constraints relative to the gain variations so that eigenvalues and eigenvectors are exactly assigned (see `fb_tun`).

Synopsis.
```
CSTR = cstr_eig(CSTR,sys,pol[,key,def_pb[,X1,X2[,X3]]],
                                        delta,KO)
```

Description. First, read page 211. The constraints relative to the gain variation k are such that KO + k assigns the eigenvalues pol(i) with eigenvectors defined together by key(i) and def_pb(:,i) (and possibly X1,X2,X3). This function belongs to the tuning sub-toolbox.

Input arguments.

CSTR   Matrix of existing constraints when cstr_eig is called.

sys,pol,key,def_pb,X1,X2,X3   Are the input arguments of defin_vw, see page 195.

delta   Such that size(delta) = size(pol). delta(i) (positive and small with respect to 1) indicates the precision of the assignment of pol(i) (exact assignment if delta(i) = 0).

KO   Initial proportional gain.

Output argument.

CSTR   Used for communication between RMCT functions.

See also: fb_tun, cstr_dp, cstr_qud, cstr_ev, cstr_k

Example. We want to assign all the poles in the quadrilateral Quad and the eigenvector relative to $-2 + i$ to have its 2nd and 3rd entries equal to zero. First, an initial gain that satisfies the eigenvector assignment is computed.
```
randn('seed',0); rand('seed',0);
sys = rss(5,3,3);
KO  = fb_prop(sys,0,-2+i,'z',[2;3]);
```
The following tuning loop preserves this assignment and shifts the other poles into the considered quadrilateral.
```
Quad = [-10+10*i -1+1*j;-10 -1];
for ii = 1:20;
   CSTR = cstr_eig([],sys,-2+i,'z',[2;3],0,KO);
   lam  = eig_fb(sys,KO);
   CSTR = cstr_qud(CSTR,sys,Quad,lam,0.01,KO);
   k = fb_tun(CSTR,[]);
   KO = KO + k;
end;
```

In order to check the results, the assigned eigenstructure is computed as follows [L,V] = eig_fb(sys,KO). The assigned eigenvalues and the eigenvector corresponding to $-2+i$ are (second and third entries equal to zero as expected):

```
>> disp([L V(:,3)])
```

```
   -3.9269 + 4.6259i   -0.1656 + 0.0395i
   -3.9269 - 4.6259i    0.0000 - 0.0000i
   -2.0000 + 1.0000i   -0.0000 + 0.0000i
   -2.0000 - 1.0000i   -0.7194 + 0.0129i
   -4.0795             -0.6585 + 0.1405i
```

Discussion. This function is very similar to **eig_cstr**. Its main specificity is that, belonging to the tuning sub-toolbox, the constraints that are defined are relative to the *gain variation* instead of relative to the gain.

In some specific cases, a nonzero input argument **delta** permits us to perform approximate assignments (**delta** must be chosen by trials and errors (each entry $\ll 0.1$)).

Note that eigenstructure assignments, using **cstr_eig**, usually leads to large feedback gain variations, so, when these constraints are combined with first order constraints, some care must be exercised. In principle, if tuning is initialized by a gain that satisfies the assignments defined by **cstr_eig**, no problem will be encountered (this assignment will be preserved during tuning iterations).

3.2.13   Function: CSTR_EV

Purpose. Computes allowable variations of the gain so that **lambda** remains in the closed-loop spectrum. Can also be used for pole assignment. This function belongs to the tuning sub-toolbox (see **fb_tun**, page 211).

Synopsis.

```
CSTR = cstr_ev(CSTR,sys,lambda,KO)
```

Input arguments.

CSTR   Matrix of existing constraints when **cstr_ev** is called.

sys   LTI system (see **ss.m**).

lambda   Vector of the eigenvalues expected to remain in the closed-loop spectrum.

KO   Initial proportional gain.

Output argument.

CSTR   Used for communication between RMCT functions.

See also: fb_tun, cstr_dp, cstr_qud, cstr_eig, cstr_k

Example: Although this function was written in order to preserve assigned eigenvalues, it can also be used for pole assignment (but convergence is sometimes erratic). In the following example we assign the following eigenvalues $\{-4 + 2j, -4 - 2j, -2, -2 + 2j, -2 - 2j\}$.

```
randn('seed',0); rand('seed',0);
sys = rss(5,3,3); KO = zeros(3,3);
lam0 = [-4+i*2;-2;-2+2*i];
for ii = 1:100;
   CSTR = cstr_ev([],sys,lam0,KO);
   k = fb_tun(CSTR,[]);
   KO = KO + k;
end;
```

After 100 iterationst the result is:

```
>>  disp([eig(sys) eig_fb(sys,KO)])

 -3.2058 + 5.6151i   -1.9884 + 1.9832i
 -3.2058 - 5.6151i   -1.9884 - 1.9832i
 -5.4529             -3.9848 + 1.9991i
 -0.4983             -3.9848 - 1.9991i
 -1.0608             -2.0000
```

Discussion. The following result is used (see (1.7)): $\lambda \in \sigma(A + BKC)$ if and only if

$$\det(KCV(\lambda) - W(\lambda)) = 0$$

where $V(\lambda)$ and $W(\lambda)$ are defined by

$$\begin{bmatrix} A - \lambda I & B \end{bmatrix} \begin{bmatrix} V(\lambda) \\ W(\lambda) \end{bmatrix} = 0$$

So, the variation of K is computed in order to shift to the origin the minimum singular value of $KCV(\lambda) - W(\lambda)$. For that purpose we use Lemma 1.2.2 (page 34) applied to the matrix $(KCV(\lambda) - W(\lambda))'(KCV(\lambda) - W(\lambda))$ (first order approximation). The software considers $D \neq 0$.

3.2.14   Function: CSTR_INI

Purpose. Initializes the matrix CSTR with no constraint except the size of the proportional feedback. This function belongs to the tuning sub-toolbox (see fb_tun, page 211).

Synopsis.
```
CSTR = cstr_ini(m,p)
```

Input arguments.

m, p   Number of inputs, number of outputs.

See also: cstr_dp, cstr_eig, cstr_ev, cstr_qud

3.2.15   Function: CSTR_K

Purpose. Defines constraints relative to a proportional gain to be tuned (see fb_tun).

Synopsis.
```
CSTR = cstr_k(CSTR,Kmin,Kmax[,K0[,nocomm]]])
```

Description. First, read page 211. Let K0 denote a given feedback gain and k denote a small variation of K0. The following linear element-wise constraints can be defined:

- Kmin(i,j) <= K0(i,j) + k(i,j) <= Kmax(i,j).

- Kmax(i,j) = Kmin(i,j) for an equality constraint, we shall have K0(i,j) + k(i,j) = Kmin(i,j).

- Kmax(i,j) < Kmin(i,j) for no constraint on k(i,j).

The input arguments must be chosen such that large variations of the gain are avoided. This function belongs to the tuning sub-toolbox.

Input arguments.

CSTR   Matrix of existing constraints when cstr_k is called.

Kmin   Matrix such that size(Kmin) = size(K0).

Kmax   Matrix such that size(Kmax) = size(K0).

K0   Initial proportional gain (= 0*Kmin by default).

`nocomm`   No comment if `nocomm == 0`.

Output argument.

`CSTR`   Used for communication between RMCT functions.

See also: `fb_tun, cstr_dp, cstr_eig, cstr_ev, cstr_ini`

Example: Consider a gain for a system having 2 inputs and 3 outputs. We want the variation ΔK of this gain to satisfy $-2 < \Delta K(1,1) < 2$; $\Delta K(2,3) = 0$ and $\Delta K(1,3) < 3$:

```
Kmin = [-2    1  -inf ;  1    0    1 ]
Kmax = [ 2   -1     3 ; -1    0   -1 ]
CSTR = cstr_k([],Kmin,Kmax,zeros(2,3),1);
```

Note that conditions of the form $1 < \Delta K(i,j) < -1$ are ignored.

Discussion. In most cases, this function must be invoked with `K0 = 0`. It is also possible to structure the global variation $K + \Delta K$, but in this case the designer must check that the corresponding conditions will not induce large variations of ΔK. If such global constraints are required, it is recommended to initialize the iterations by an initial gain satisfying the global constraints (see page 86). A similar problem concerning the function `cstr_eig` is treated in § 3.2.12. See also page 295 for the the structure of the resulting matrix `CSTR`.

3.2.16   Function: CSTR_QUD

Purpose. Defines linear constraints for pole shifting into a quadrilateral (see `fb_tun`).

Synopsis.
`CSTR = cstr_qud(CSTR,sys,Quad,lambda,speed,K0[,target])`

Description. First, read page 211. The quadrilateral is defined by `Quad`. Only the closed-loop poles the closest to the entries of the vector `lambda` (if they are outside the quadrilateral) are considered for defining pole shifting constraints. For each treated eigenvalue, three constraints are defined as illustrated in Figure 1.15, page 85. This function belongs to the tuning sub-toolbox.

Input arguments.

`CSTR`   Matrix of existing constraints when `cstr_qud` is called.

sys   LTI system (see **ss.m**).

Quad   Defines a quadrilateral into which the closed-loop poles the closest to the entries of **lambda** must be shifted. **Quad** is a 2 by 2 complex
matrix defining the corners of the quadrilateral (which must be inside
the upper half complex plane including the real axis).

lambda   Vector of the eigenvalues (with positive imaginary part) that
must be shifted into the quadrilateral. If all eigenvalues are to be
shifted set **lambda = eig_fb(sys,KO)**. See also **sort_ev**.

speed   Is a scalar number s.t. 0 < **speed** <= 0.1: **speed** must be
large enough to obtain fast convergence and small enough to avoid
non-convergence.

KO   Initial proportional gain.

target   Complex number: pole target inside the quadrilateral (point
Q of Figure 1.15).

Output argument.

CSTR Used for communications between RMCT functions.

See also: fb_tun, sort_ev, dist_qud

Example: All the eigenvalues are shifted into the trapezium with upper
left and right corners $-10 + 10i$ and $-1 + j$ and lower left and right
corners -10 and -1. Use "trials and errors" for the choice of the sixth
input argument (**speed**). First, is defined the system and criterion:

```
randn('seed',0); rand('seed',0);
sys = rss(10,3,3);
KO = zeros(3,3);
Quad = [-10+10*i -1+1*j;-10 -1];
CRIT = crit_k(ones(3,3),zeros(3,3));
```
The tuning loop is:
```
for ii = 1:60;
    lam = eig_fb(sys,KO);
    CSTR = cstr_qud([],sys,Quad,lam,0.01,KO);
    k = fb_tun(CSTR,CRIT);
    KO = KO + k;
end;
```
and the results are (**[eig_fb(sys) eig_fb(sys,KO)]**):
```
-1.8717 + 4.2640i  -4.3049
```

```
-1.8717 - 4.2640i   -2.4739 + 2.9886i
-3.2058 + 3.2339i   -2.4739 - 2.9886i
-3.2058 - 3.2339i   -2.4571 + 2.4050i
-3.4763             -2.4571 - 2.4050i
-2.3909             -2.3384 + 0.9544i
-1.3022             -2.3384 - 0.9544i
-0.2355             -1.0303 + 0.1174i
-0.5278             -1.0303 - 0.1174i
-0.4960             -0.9222
```

All the eigenvalues have almost reached the trapezium. Other examples: see page 145 (and page 92).

Discussion. See Figures 1.15 and 1.16, page 85. See also page 295 for the structure of the resulting matrix CSTR.

3.2.17   Function: **DEFIN_VW**

Purpose. Computes the right eigenvectors that can be assigned for given desired closed-loop eigenvalues. Several options are offered.

Synopsis.
```
[v,w,cv] = defin_vw(sys,pol,key,def_pb[,cc,dd[,ee]]);
[v,w,cv] = defin_vw(sys,pol,'p',def_pb,fb0);
[v,w,cv] = defin_vw(sys,pol,'p',def_pb,K,'s');
[v,w,cv] = defin_vw(sys,pol); <- single-input system
```

Description. The options are:

- some entries of eigenvectors set to zero $\rightarrow$ 'z'

- minimum energy assignment $\rightarrow$ 'm'

- projection of eigenvectors $\rightarrow$ 'p'

- eigenvector in a null space $\rightarrow$ 'n'

- least squares with respect to a desired vector $\rightarrow$ 'v'

- assignment of more than p eigenvalues. $\rightarrow$ 'e'

(p is the number of measurements). With options 'z' and 'n', the problem can be under or over specified. In the first case, a feasible choice is made at random, in the second case, least squares is used.

Input arguments.

sys   LTI system (see `ss.m`).

pol   Vector (of length q) of assigned eigenvalues. Do not repeat conjugate values.

key   Vector (of length q) defining options for assignment: `key(i)` is one of the strings `'z'`, `'n'`, `'v'`, `'p'`, `'m'`, `'e'`. If all assignments are made using the same option, `key` can be one of the 1 by 1 strings listed above (except the last one).

def_pb   Matrix having q columns. Non significant entries should be set to zero. Each column has a different meaning depending on the corresponding option, as detailed below.

cc,dd,ee   These three matrices are useful only if `key(i)` = `'n'` for some i.

fb0   This feedback gain argument is relevant only if `key` = `'p'`. Projection of eigenvectors of `sys` controlled by `fb0`. If the reference feedback is a state feedback K, replace `fb0` by `K,'s'`.

key(i)   == `'z'`:   `def_pb(:,i)` contains the indices of the entries of the eigenvector corresponding to `pol(i)` that must be set to zero. Indices larger than the number of states refer to the entries of the input direction vector.

== `'n'`:   `def_pb(:,i)`: The vectors vi (right eigenvector) and wi (input direction) corresponding to `pol(i)` will be such that `cc(def_pb(:,i),:)*vi + dd(def_pb(:,i),:)*wi =` `ee(def_pb(:,i),i)`. Usually ee = 0. Default cc = `sys.c`, dd = `sys.d`, ee = 0. See page 31 for an example.

== `'v'`:   `def_pb(:,i)` is a $n \times 1$ vector. (n number of states). `def_pb(:,i)` is the desired (possibly complex) ith eigenvector (it is assigned by least squares minimization).

== `'p'`: for projection of open-loop eigenvector. Three cases

1- one open-loop real (resp. nonreal) pole replaced by a closed-loop real (resp. nonreal) one: `pol(i)` is the assigned (closed-loop) eigenvalue `def_pb(1,i)` is the concerned open-loop eigenvalue.

2- two open-loop real poles `OL_pole_1` and `OL_pole_2` replaced by one closed-loop nonreal pole `CL_pole`. Then, `pol(i)` = `CL_pole` and `def_pb(1,i)` = `OL_pole_1` + j `OL_pole_2`, `def_pb(2,i)` = 1.

3- one open-loop nonreal `OL_pole` replaced by two closed-loop real `CL_pole_1` and `CL_pole_2`. Then, `pol(i)` = `CL_pole_1` + j `CL_pole_2`, `def_pb(1,i)` = `OL_pole`, `def_pb(2,i)` = 2.

== 'm': minimum energy assignment `def_pb(:,i)` contains the open-loop pole that must me shifted to `pol(i)` with "minimum energy".

== 'e': the pth eigenvector assigns r other eigenvalues that are specified in `def_pb(1:r,p)`, `pol(p)` must be real $r < m$ (m = number of inputs, p = number of outputs).

`key` == 'p'   Offers the possibility of selecting projection of closed-loop eigenvectors.

Output arguments.

`v,w,cv`   Assignable eigenvectors, input directions and vectors cv: cv represents `sys.c * v + sys.d * w` (see Equation (1.30), page 28).

See also: `fb_prop, ff_assgn, ob_gene, eig_cstr, cstr_eig`

Example 1. Although this function is not directly used (it is invoked by other functions), in the special case of multi-model assignment with proportional gain, using `defin_vw` is much simpler than using `fb_dyn`. See page 140 for an illustrative example.

Example 2. Illustration of a multi-objective eigenvector selection.

```
pol = [-0.2 -0.3 -0.2+0.3*i -0.5 -1.5];
key = [ 'z'   'n'      'v'        'p'   'm'];
```

Now, let us describe the constraint defined by the matrix `def_pb`. The first column defines the entries of the eigenvector corresponding to -0.2 that must be equal to zero (option 'z'). Here it is the 8th one. The second column defines which rows of "C" and "D" are concerned for the null space (option 'n') in the computation of the right eigenvector corresponding to -0.3 . Here, it is the second row. The third column is the desired right eigenvector for $-0.2 + 0.3 * i$ (option 'v'). The two other rows define open-loop eigenvalues (option 'p' and 'm').

```
def_pb = [];
def_pb(1,1) = 8;
def_pb(1,2) = 2;
def_pb(1:8,3) = [0;0;0;-0.7+0.6*i;0;0;0.02;0.02+0.04*i];
def_pb(1,4) = -0.18;
def_pb(1,5) = -1.3;
```

```
[V,W,CV] = defin_vw(sys,pol,key,def_pb);
```

Discussion. Usually, this function can be ignored by the user, except for the help message which is not repeated in the functions that have the same input arguments. These functions are `fb_prop`, `eig_cstr`, `cstr_eig`, `ob_gene` and `ff_assgn`. More details on the options can found on page 21 for 'n' and 'z', page 22 for 'p' and 'v', page 24 for 'm' and page 33 for 'e'.

3.2.18   Function: DEMODATA

Purpose. Generates linear models for software demonstration.

Synopsis.
```
[sys,decrip] =  demodata(Isys)
```

Input arguments.

Isys   Lateral model of a rigid aircraft (6 states) if `Isys` = 1. Similar, but including flexibility if `Isys` = 2.

Output arguments.

sys   LTI representation of the system (see ssdata.m).

descrip   Text describing briefly the system.

See also: `rcamdata`

Discussion. See page 289.

3.2.19   Function: DFB_INS

Purpose. Computes an observer-based dynamic feedback by minimization of closed-loop eigenvalue sensitivity. The criterion that is minimized is the one of page 35. This function is a combination of `sfb_ins` and `ob_ins`.

Synopsis.
```
[fb,ff,crit] = dfb_ins(sys,polc,polo[,niter[,op_crit]]);
```

Caution. Exact pole assignment is relevant for low dimensional systems. For partial pole assignment, or in case of problems induced by weakly controllable poles, it is suggested to customize `dfb_ins` by considering directly `ob_ins` and `sfb_ins` (or to use `dfb_proj` instead of `dfb_ins`).

Input arguments.

`sys` LTI system (see `ss.m`).

`polc` State feedback closed-loop poles.

`polo` Observer poles.

`niter` Number of iterations. More precisely, `niter(1)` is the number of random trials before optimization, `niter(2)` is the number of iterations of the heuristic algorithm of §1.2.3, `niter(3)` is the number of iterations of standard optimization. If this argument is scalar, `niter` is internally replaced by `niter*[1 1 1]`. If this argument is omitted `niter` is considered by default as equal to the number of states of `sys`.

`op_crit` = 'k' for minimization of the gain norms, = 'c' for optimization of the condition numbers (default), = `[coef_k coef_c]` (two real numbers) for a linear combination of both above criteria.

Output arguments.

`fb` Dynamic full order observer-based feedback. (LTI system see ss-data.m). See Equation (1.75).

`ff` Dynamic feedforward (LTI system). See Equation (1.76).

`crit` Values of the criterion of Equation (1.45) at the various stages of the optimization (`crit(:,1)` = initial values, `crit(:,2)` = values after random trials, `crit(:,3)` = after heuristic optimization, `crit(:,4)` = optimal values). The first row concerns the state feedback design, the second one concerns the observer design.

See also: `sfb_ins`, `ob_ins`, `dfb_proj`

Example. See page 123.

3.2.20   Function: DFB2OBS

Purpose. Transforms a standard dynamic feedback into an observer-based dynamic feedback.

Synopsis.

```
[U,T,Pi,Q,K,ff,Ky,Kz] = dfb2obs(sys,fb,polref);
```

Description. If n_c is the order of the controller to be transformed, n_c closed-loop poles must be selected as observation dynamics. The resulting observer is defined by the matrices U, T, Pi (see page 47), the resulting proportional feedback is defined by Ky and Kz where the feedback is given by $u =$ Ky $y +$ Kz $\hat{z}$. If input/output behaviour is of interest, the feedforward gain ff must be added at the system / observer input. (See add_obs for connecting the observer to sys.) The feedback K = [Ky Kz] can be tuned by using fb_tun.

Input arguments.

sys   LTI system (see ss.m).

fb   Dynamic feedback gain to be transformed.

polref   Vector of closed-loop poles to be considered as observer dynamics. If polref does not correspond to closed-loop poles, observer dynamics are selected by a default procedure. For example, if polref = 0 slow poles are automatically selected, if polref = -100 it is fast poles that are selected. Do not repeat conjugate values in this argument.

Output arguments.

U,T,Pi,Q   Matrices characterizing the resulting observer.

K   Concatenated feedback matrix: K = [Ky Kz].

Ky   Part of the feedback gain relative to measurements y.

Kz   Part of the feedback gain relative to the observed signal $\hat{z}$.

ff   Dynamic feedforward gain.

See also: obs2dfb, add_obs

Example. This example concerns the transformation of a dynamic feedback of order n_c less than the number of states n. The equivalent observer is of order $n_c < n$. (There is no difference with the case $n_c > n$ leading to an observer of order n_c larger than n, see help2 dfb2obs.)

```
rand('seed',5); randn('seed',0);
sys = rss(6,4,2);
fb  = rss(5,2,4);
polref = [-3+6i -7.7 -4+0.2i];
```

```
[U,T,Pi,Q,K,ff] = dfb2obs(sys,fb,polref);
```

It remains to connect to `sys` the observer described by the output arguments of the previous command line

```
sysobs = add_obs(sys,U,T,Q);
```

For a comparison with the original system, the feedback loop is closed and the dynamic feedforward is added at the system input:

```
sysequiv = feedback(sysobs,K,1)*ff;
```

The following simulations show that from the input output point of view `sysequiv` is equivalent to the original system (`sys` controlled by `fb`).

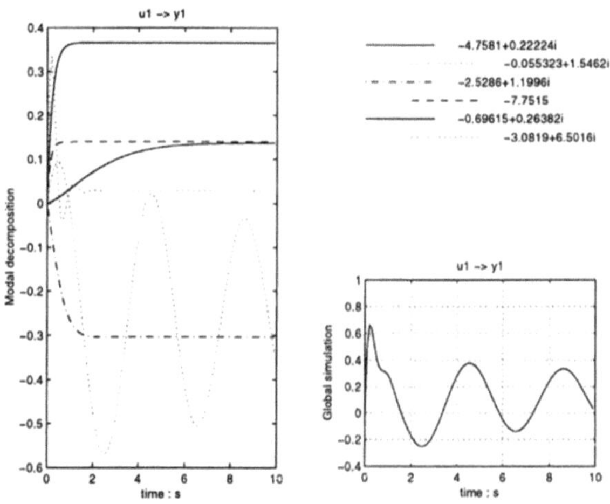

Figure 3.1. Function `dfb2obs`: Original closed-loop system

```
lsim_mod(feedback(sys,fb,1),'u1',10,0,'y1');
lsim_mod(sysequiv,'u1',10,0,'y1');
```

In the modal simulation of Figure 3.2, the four modes that seem to be dominant compensate themselves (feedforward effect), otherwise the global output (right hand side simulations) are similar.

Other example: see page 126.

Discussion. See Appendix 4, page 284.

3.2.21　Function: DFB_PROJ

Purpose. Computes an observer-based dynamic feedback by projection of the open-loop eigenstructure. The minimum damping ratio and the

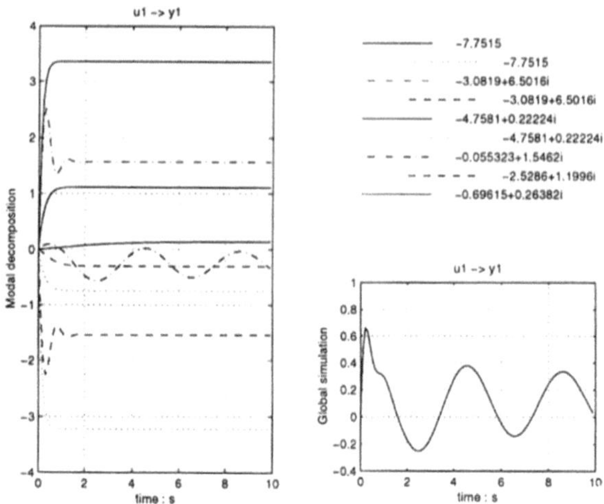

Figure 3.2. Function **dfb2obs**: Equivalent observer-based design

settling time are the design parameters. This function is a combination
of **sfb_proj** and **sob_proj**.

Synopsis.
```
[fb,ff,polc,polo] =
          dfb_proj(sys,Tr,Damp[,Tro,Dampo][,key[,C_ratio]])
```

Caution. Do not use this function if there are repeated open-loop eigenvalues.

Input arguments.

sys LTI system (see **ss.m**).

Tr Settling time for state feedback.

Damp Minimum damping (0 < Damp < 1) for state feedback.

Tro Settling time for the observer, default: Tro = Tr.

Dampo Minimum damping (0 < Dampo < 1) for the observer, default:
Dampo is equal to Damp.

key == 'p' for orthogonal projection of all eigenvectors, == 'm' for
minimum energy assignment (default).

C_ratio If the controllability degree of some eigenvalue is less than
the maximum degree of other eigenvalues divided by C_ratio, this
eigenvalue is ignored. (Default: C_ratio = 100.)

Output arguments.

fb   Feedback (LTI system, see ssdata.m).

ff   Feedforward gain ($\rightarrow$ feedback(sys,fb,1)*ff;)

polc   Poles assigned by state feedback.

polo   Poles of the observer.

See also: sfb_proj, sob_proj, dfb_ins

Example. Design of a dynamic feedback gain such that the damping ratio is larger than 0.707 and the settling time less than 15 seconds.

```
sys = rss(8,4,2);
Tr = 15; Damp = 0.707;
fb = dfb_proj(sys,Tr,Damp);
eig_fb(sys,fb);
```

Other example: see page 121.

Discussion. See §2.2.1, page 118.

3.2.22   Function: DIST_QUD

Purpose. Computes the distance of a set of poles from a quadrilateral (see cstr_qud).

Synopsis.

```
[dist,dist_1] = dist_qud(sys,Quad,lambda,K0)
```

Description. Only the closed-loop poles recognized in the entries of lambda are considered. This function can be used for a direct optimization (using the Optimization Toolbox instead of cstr_qud and fb_tun) or for storing the best result computed within a tuning loop based on the use of cstr_qud and fb_tun.

Input arguments.

sys   LTI system (see ss.m).

Quad   Defines a quadrilateral from which the distances of closed-loop poles must be computed (see cstr_qud).

lambda   Vector of the eigenvalues (with positive imaginary part) considered for distance computation.

K0   Proportional gain.

Output arguments.

dist_l   Vector of the distances of each pole.

dist   Maximum value of dist_l.

See also: `cstr_qud`, `sort_ev`

3.2.23   Function: DP_CSTR

Purpose. Multi-model and multivariable generalized "phase control" (see `fb_dyn`).

Synopsis.

CSTR = dp_cstr(CSTR,sys,pol,fld1,fld2,fld3);

Description. First, read page 208. `dp_cstr` defines first order constraints for pole shifting in a form that can be treated by `fb_dyn`. The considered constraints are relative to the real and imaginary parts variations. The variations can also be constrained on a line having a given slope. This function belongs to the multi-model sub-toolbox.

Input arguments.

CSTR   Set of existing constraints when `dp_cstr` is called. See `help1` `fb_dyn`.

sys   Considered system (see `ss.m`).

lam, fld1, fld2 ,fld3   are vectors of same length defining the desired constraints for pole shifting (at the departing points of a root locus).

pol   Vector of the poles of `sys` to be shifted.

fld1 Vector containing 'r', 'i' or 's' (for real part, imaginary part and slope).

fld2   Vector containing '=' , '<' or '>'.

fld3   Vector containing numerical values. Example:
```
lam  = [lam1; lam1;lam2];
fld1 = ['r' ; 'i' ; 's'];
fld2 = ['<' ; '=' ; '='];
fld3 = [-0.1; 0 ; 1 ];
```

means that the variations of the root locus around `lam1` and `lam2` are such that
real part of `d(lam1)` less than -0.1,
imaginary part of `d(lam1)` equal to zero
imaginary part of `d(lam2)` divided by its real part equal to 1.

Output argument.

CSTR Original plus newly computed constraints. To be used as input argument of **fb_dyn**, **eig_cstr**, **dp_cstr**, **add_cstr**.

See also: `fb_dyn, eig_cstr, add_cstr`

Example. See page 76. This example illustrates multi-model "phase control".

Discussion. See §1.3.5, page 70. The theory behind this function is given on page 259 (from [Magni et al., 1997a]). Page 295 is given the structure of the resulting matrix CSTR.

3.2.24 Function: DYN2STA

Purpose. Transforms a dynamic feedback relative to a system into an equivalent proportional gain relative to an extended system (extended by **add_dyn**).

Synopsis.
```
kprop = dyn2sta(fbdyn,dyn);
```

Description. See **add_dyn** page 175.

Input arguments.

fbdyn Dynamic feedback gain in LTI form (see **ss.m**).

dyn Vector that is jointly used as input argument of **add_dyn**.

Output argument.

kprop Equivalent proportional feedback.

See also: `add_dyn, sta2dyn`

Example. First, a dynamic LQG feedback gain **fbdyn** is designed relative to the system **sys**. Then, this dynamic gain is transformed using dyn2sta into a proportional feedback gain k1 relative to the extended system sysdyn obtained from sys using add_dyn.

System and dynamic feedback:

```
sys = rss(5,2,2); [a,b,c,d] = ssdata(sys);
[ac,bc,cc,dc] = lqg(a,b,c,d,eye(7,7),eye(7,7));
fbdyn = ss(ac,-bc,cc,-dc);
```

Dynamic extension:

```
dyn = [-1;-2+i;-2;-1];
sysdyn = add_dyn(sys,dyn);
k1 =  dyn2sta(fbdyn,dyn);
```

Results (both spectra are equal).

```
[eig_fb(sysdyn,k1) eig_fb(sys,fbdyn)]
```

Other examples. See pages 46 and 145. The example of page 46 illustrates the use of a dynamic extension for exact pole assignment by output feedback. The one of page 145 illustrates dynamic feedback gain tuning using fb_tun.

Discussion. See page 46.

3.2.25   Function: EIG_CSTR

Purpose. Defines eigenstructure assignment constraints in a form that can be treated by fb_dyn. This function belongs to the multi-model sub-toolbox (see page 208).

Synopsis.

```
CSTR = eig_cstr(CSTR,sys,pol,key,def_pb[,cc,dd[,ee]]);
CSTR = eig_cstr(CSTR,sys,pol,'p',def_pb,fb0)  <- output fb
CSTR = eig_cstr(CSTR,sys,pol,'p',def_pb,K,'s')<- state fb
```

Input arguments.

CSTR   Set of existing constraints when eig_cstr is called. See help1 fb_dyn.

All other input arguments are defined in help1 defin_vw, see page 195.

Output argument.

CSTR   Original plus newly computed constraints. To be used as input argument of fb_dyn, dp_cstr, add_cstr.

See also: defin_vw, fb_dyn, dp_cstr, add_cstr, str_cstr

Examples. See pages 74, 142. These examples illustrate the use of `eig_cstr` for multi-model dynamic feedback design.

Discussion. This function permits the designer to perform multi-model assignment in an algebraic way. However, its use requires much care in order to avoid incompatibility within multi-model requirements. In case of problem, the user must read carefully the recommendations listed in §2.1, page 109, with a special attention to those relative to the multi-model sub-toolbox. See page 195 for details on the input arguments (do not use option `'e'`). The theory behind `eig_cstr` is briefly presented in Lemma 1.3.1 page 65, details can be found in Chapter 4 and in [Magni et al., 1998, Magni, 1999], some applications can be found in [Le Gorrec et al., 1997, Le Gorrec et al., 1998b]. (The structure of the resulting matrix CSTR is given on page 295).

3.2.26   Function: EIG_FB

Purpose. Computes closed-loop eigenstructure.

Synopsis.
```
[D,V,W,U,T] = eig_fb(a,b,c,d[,ac,bc,cc],dc)
[D,V,W,U,T] = eig_fb(sys,fb)
[D,V,W,U,T] = eig_fb(sys,k[,opt])
```

Input arguments.

a,b,c,d   System. c=1 means that c is the identity matrix. d=0 means that d is the zero matrix with appropriate size.

sys   Alternative definition of the system (sys = ss(a,b,c,d)).

ac,bc,cc,dc   Dynamic output feedback gain. dc=0 means that dc is the zero matrix with appropriate size.

fb   Alternative definition of the feedback gain (fb = ss(ac,bc,cc,dc)).

k   Proportional output feedback.

opt   String: opt = `'s'` if k is a state feedback.

Output arguments.

D   Vector of eigenvalues.

V,W   Matrices of right eigenvectors, input directions.

U,T   Matrices of left eigenvectors, output directions.

See also: eig, choi_ev

Examples. Numerous examples are given in this manual.

Discussion. In the proportional gain case, the matrices V,W,U,T are as defined on page 15. In the dynamic gain case, eigenstructure is computed as discussed in the comment of page 65.

3.2.27   Function: FB_DYN

Purpose. Computes a dynamic, multi-model, structured feedback. It is the main function of the multi-model sub-toolbox. All other functions of this sub-toolbox are used in order to generate both main input arguments of fb_dyn.

Synopsis.
```
[fb,k,dyn,ksi,valcrit] = fb_dyn(CSTR[,CRIT[,tol]]);
```

Description. The use of a dynamic feedback permits the designer to assign more than p (number of measurement) eigenvectors (only p assignments when fb_prop is used). List of the design objectives that are possible to meet using fb_dyn:

- multi-model eigenstructure (eigenvalues & eigenvectors) assignment (see eig_cstr),

- multi-model "phase control" (see dp_cstr),

- use of structured a feedback gain (see str_cstr),

- frequency domain gain templates (see add_cstr).

Furthermore, the distance with a reference dynamic feedback can be minimized (see ktf_crit).

Input arguments.

CSTR   Constraints relative to the gain CSTR is built by the functions str_cstr, eig_cstr, dp_cstr, add_cstr.

CRIT   Is a criterion corresponding to the distance between a reference feedback and the feedback to be computed. This matrix is generated by ktf_crit. By default the reference feedback is the zero-feedback.

`tol` Tolerance for controller order reduction, default value: `100*eps` (see `help minreal`).

Output arguments.

`fb` Is the resulting minimal state-space form of the feedback (see `ssdata.m`).

`k,dyn` See `help1 add_dyn`. This function generates an augmented system that is controlled by the constant feedback gain `k`.

`ksi` Coefficients of the transfer function matrix form of the gain, see `fb_view`.

`valcrit` Criterion value at the optimum.

See also: `fb_prop, fb_tun`

Examples. See pages 76, 142 and 153. These examples concern respectively the following problems: multi-model phase control with frequency domain constraints, multi-model eigenstructure assignment and single-model phase control.

Discussion. All the potentialities and limitations of `fb_dyn` are discussed in §1.3 (page 63). Details on theory can be found in Chapter 4 and in [Magni, 1999, Magni et al., 1998], applications can be found in [Chiappa et al., 1998, Le Gorrec et al., 1997, Le Gorrec et al., 1998b, Magni et al., 1997a].

3.2.28   Function: FB_PROP

Purpose. Assigns eigenstructure (eigenvalues and eigenvectors) by proportional (state or output) feedback.

Synopsis.
```
k = fb_prop(sys,k0,pol,key,def_pb[,cc,dd[,ee]]);
k = fb_prop(sys,k0,pol,'p',def_pb,fb0);
k = fb_prop(sys,k0,pol,'p',def_pb,K,'s');
k = fb_prop(sys,k0,pol); <- single-input system
```

Description. See help `defin_vw`.

Input arguments. All input arguments are similar to those of the function `defin_vw` except:

k0   If the problem is under-specified, k is computed as being the feasible gain the closest to k0. (k0=0 means that k0 is the zero matrix with appropriate size).

Output argument.
 k   Feedback gain.

See also: defin_vw, fb_dyn, fb_tun

Example 1. Assignment of $\{-0.1 + 0.1j, -0.1 - 0.1j, -2, -4\}$. The eigenvectors associated with $-0.1 + 0.1j$ and its conjugate value are obtained by projection of the open-loop eigenvectors associated to both open-loop eigenvalues $\{-1, -0.5\}$. Assignment of -2 and -4 are made with minimum energy from $+2$ and -4. The open-loop eigenvalues -3 and -6 are expected to be moderately perturbed by the designed feedback on account of the use of projection and minimum energy ideas.

```
a=diag([-1;2;-3;-4;-0.5;-6]);
b=[eye(2,2);eye(2,2);eye(2,2)];
c=[eye(4,4) ones(4,2)];
sys = ss(a,b,c,0);


key  = [ 'p'                    'm'     'm'   ];
pol  = [ -.1+.1*i               -2      -4    ];
def_pb=[[(-1)+i*(-0.5);1] [2;0]    [-4;0]];
k=fb_prop(sys,0,pol,key,def_pb)
```

Example 2. Assignment of $\{-0.6, -1.1, -2, -5.9, -3.9\}$. The assignment of the 3 first eigenvalues is made respectively by (projection, projection, minimum energy) from open-loop eigenstructure $(-0.5, -1, +2)$. The assignment of -5.9 is such that an additional eigenvalue -3.9 is also assigned.

```
key     = [ 'p'       'p'       'm'      'e' ];
pol     = [ -0.6      -1.1      -2       -5.9];
def_pb  = [ -0.5      -1        2        -3.9];
k=fb_prop(sys,0,pol,key,def_pb)
```

Example 3. Assignment of $\{-1.1, -0.6, -2, -5.9\}$. The 2nd entry of the eigenvector associated with -1.1 must be zero. The 1st entry of the input direction associated with -0.6 must be zero. Minimum energy assignment for -2 and -5.9 from open-loop eigenvalues $+2$ and -6.

```
key     = [ 'z'       'z'       'm'      'm'  ];
pol     = [ -1.1      -0.6      -2       -5.9 ];
def_pb  = [  2         7         2       -6   ];
```

```
k=fb_prop(sys,0,pol,key,def_pb)
```

Discussion. This function solves the problems discussed in §1.2.1, page 28. The basic references are [Kimura, 1975, Moore, 1976]. One of the most well-known application can be found in [Farineau, 1989].

3.2.29　Function: **FB_TUN**

Purpose. Computes the gain variation (of a proportional gain) which minimizes a criterion and satisfies a set of linear constraints built by the other functions of the tuning sub-toolbox. It is the main function of the the tuning sub-toolbox.

Synopsis.

```
[k,flg] = fb_tun(CSTR,CRIT[,nocomm])
```

Description. The function **fb_tun** computes gain variations, so, it is essential to use it within an iterative tuning procedure (to be written by the user). List of the design objectives that can be met:

- multi-model-tuning (all constraints can be computed with respect to more than one model),

- tuning of closed-loop poles (see **cstr_dp**, **cstr_qud**, **cstr_ev**),

- eigenvector assignment (see **cstr_eig**),

- eigenvector tuning (see **comp_dv**),

- gain structure (see **cstr_k**).

An extension to dynamic feedback tuning is possible using **add_dyn**, **dyn2sta** and **sta2dyn**, or using **add_obs**, **dfb2obs** and **obs2dfb**

Note that if there are too many equality constraints, the inequality constraints are ignored and a least squares technique is used for equalities.

Input arguments.

CRIT　Built by **crit_k**, **crit_ctr**. CRIT = [] means that the gain variation (**trace(k'*k)**) is minimized.

CSTR　Built by **cstr_k**, **cstr_dp**, **cstr_qud**, **cstr_eig** and **cstr_ev**.

nocomm　No comment if **nocomm** == 0.

Output arguments.

k   Computed gain variation.

flg   If flg = 'ok' the results are reliable.

See also: quadprog (Optimization Toolbox), fb_dyn, fb_prop

Examples. See pages 89, 92, 145, 186, 188, 193....

Discussion. All the potentialities of fb_tun are presented in §1.4 (from page 79). See [Magni and Manouan, 1994] for the theory and [Livet, 1995] for applications.

3.2.30   Function: FB_VIEW

Purpose. Displays to the screen the transfer function matrix of the gain computed by fb_dyn.

Synopsis.
```
fb_view(CSTR,ksi[,entries])
[const,num,den] = fb_view(CSTR,ksi,[i j])
```

Input arguments.

CSTR   Constraints that were used as input argument of fb_dyn.

ksi   Fourth output argument of fb_dyn.

entries   This is a q by 2 matrix, each of its row defines one of the q gain entries that is to be displayed. Default: all entries are displayed.

Output arguments. The output arguments are generated only if the third input argument is of the form [i j] (i and j integers)

const   Direct transmission coefficient of gain entry (i,j).

num   Numerator coefficients (strictly proper part) of gain entry (i,j).

den   Denominator coefficients of gain entry (i,j).

See also: str_view, fb_dyn

Discussion. The gain structure is defined in (1.85), page 65.

3.2.31   Function: **FF_ASSGN**

Purpose. Multivariable pole / zero cancellation with "assignment" of the right eigenvectors corresponding to the new poles.

Synopsis.
```
dff=ff_assgn(sys,k,pol_old,zer,pol,key,def_pb[,cc,dd[,ee]]);
dff=ff_assgn(sys,k,pol_old,zer,pol); <- single-input case
```

Description. This function computes a feedforward gain which cancels some poles (of **sys** controlled by **k**) by assigning zeros at the same location. The cancelled poles (`pol_old`) are replaced by new ones (`pol`) that are chosen by the designer. If the zeros (**zer**) coincide exactly with `pol_old`, the cancellation is exact otherwise it is made approximately. Note that:

- the three vectors `pol_old`, **zer** and **pol** must have entries of the same nature (*i.e.*, real or nonreal, without repeated conjugate values).

- **zer** must be close enough to `pol_old`.

- the entries of `pol_old` must almost coincide with some entries of `eig_fb(sys,k)`.

Input arguments.

sys   LTI system (see **ss.m**).

k   Proportional feedback (k=0 means that k is the zero matrix with appropriate size) or LTI dynamic feedback. The function closes the feedback loop before cancellations.

pol_old   Is a q by 1 vector of cancelled eigenvalues.

zer   Is a q by 1 vector of zeros close to `pol_old`.

pol,key,def_pb   Similar to input arguments of **defin_vw**.

Output argument.

dff   Dynamic feedforward (LTI system, see **ssdata.m**).

See also: `fb_prop`, `defin_vw`, `ff_stat`

Example. It is expected to replace the poles the closest to $-0.1 + 1j$ and to -2 by poles at $-1 + 1j$ and -5 (`pol_o = [-0.1+1*j;-2]` and `pol_n = [-1+1*j;-5]`). The poles at $-0.1 + 1j$ is almost cancelled by a zero at $-0.2 + 1j$ and the pole at -2 is exactly cancelled (`zer = [-0.2+1*j;-2]`). The eigenvector corresponding to the new pole $-1+1j$ must have its first and second entries equal to zero, the other eigenvector must have its first and fourth entries equal to zero (option `'z'` for both assignments and `d_p = [[1;2] [1;4]]`).

```
a=[-1 0 0 0;0 -.1 -1 0;0 1 -.1 0; 0 0 0 -2];
b=[0 0 1;1 1 1;0 1 0;1 0 1];
sys = ss(a,b,eye(size(a)),0);
key = 'z';
d_p = [[1;2]   [1;4]];
pol_o = [-0.1+1*j;-2];
zer   = [-0.2+1*j;-2];
pol_n = [-1+1*j;-5];
dff = ff_assgn(sys,0,pol_o,zer,pol_n,key,d_p);
```

The results can be checked by computing the eigenvectors after connection of `dff` and `sys` in series. On the diagonal basis, the entries of the "B-matrix" in front of eigenvalues of `pol_o` must be close to zero (poles made almost uncontrollable).

Other examples can be found on pages 44 and 134.

Discussion. The algorithm behind is described in Lemma 1.2.3, page 43. The original version of this result can be found in [Magni, 1987].

3.2.32   Function: FF_STAT

Purpose. Computes a feedforward gain that eliminates cross-coupling in the steady-state response.

Synopsis.
```
[ff,dh] = ff_stat(sys,k,def_pb[,cc[,dd]]);
```

Description. Let `z` be the vector of regulated outputs (see "inputs arguments" below) and `z_r` the vector of reference inputs (in `u = k y + dh z_r`). The function `ff_stat` renders the steady state gain between `z_r` and `z` equal to the identity matrix. Note that the number of inputs of the system `sys` must be larger than or equal to the number of regulated outputs.

Input arguments.

sys   LTI system (see **ss.m**)

k   Proportional feedback gain (k=0 means that k is the zero matrix with appropriate size) or LTI dynamic feedback. The function ff_stat closes internally the feedback loop before computing the feedforward gain.

def_pb   Selects the rows in cc and dd for defining the regulated outputs. The regulated outputs are given by z = cc(def_pb,:)*x + dd(def_p,:)*u.

cc,dd   By default cc = sys.c and dd = sys.d.

Output arguments.

ff   Constant feedforward in LTI form (see ssdata.m).

dh   Constant feedforward in matrix form.

See also: ff_assgn, dcgain

Examples. This function is used several times in this manual, see for example pages 44, 131. Explanations can be found on page 38.

3.2.33   Function: KTF_CRIT

Purpose. Generates a criterion relative to the distance between a reference dynamic feedback gain and the feedback gain to be computed by fb_dyn.

Synopsis.
CRIT = ktf_crit(CSTR,fb0,omega[,weight[,CRITini]]);

Description. First, read page 208. Let $K_0(s)$ denote the reference feedback and $K(s)$ the feedback being computed. The considered criterion is

$$J = \sum_i \alpha_i \, \mathrm{trace}(K(\omega(i)) - K_0(\omega(i))) \, (K(\omega(i)) - K_0(\omega(i)))'$$

This function belongs to the multi-model sub-toolbox.

Note that this criterion is naturally badly conditioned when the values in omega are chosen far from the unity (see page 254). Conditioning is even worst when these values are outside the frequency bandwidth of the system. The function that minimizes this criterion (fb_dyn) gives a warning message when the criterion is too badly conditioned and add internally a term for regularisation.

Input arguments.

CSTR   Set of existing constraints when ktf_crit is called. CSTR must contain the constraints relative to the feedback structure, *i.e.*, it is assumed that the functions str_cstr has already initialized this matrix. Otherwise if CSTR = [], the feedback is considered as being proportional.

fb0   Is a reference feedback gain (LTI system see ss.m or matrix).

omega   Is the set of complex values where the distance between the reference and computed gains is considered. The minimum length of omega should be equal to the maximum of the denominator degrees plus one. If all entries of omega are real, it is assumed that this vector contains frequencies *i.e.*, omega is internally multiplied by j $(= \sqrt{-1})$.

weight   Vector of weighting coefficients (denoted α_i in the formula defining the criterion J). Same length as omega.

CRITini   The criterion defined here is added to the one in CRITini (which was generated by calling ktf_crit). Useful for example if two reference feedback gains are to be considered in distinct frequency intervals.

Output argument.

CRIT   Form of the criterion to be used as input argument of fb_dyn or ktf_crit.

See also: fb_dyn

Examples. See pages 74, and 143.

Discussion. The criterion is defined by (1.86), page 66. A closed form, written in terms of the gain coefficients (1.85), can be found on page 253 (from [Magni, 1999]). On account of the aforementioned natural bad conditioning of the criterion, it is recommended to limit to four or five the denominator degrees of the gain coefficients.

3.2.34   Function: KTF_CSTR

Purpose. Can be ignored by the toolbox user. Transforms gain structure requirements relative to a dynamic feedback (in transfer function matrix form) into constraints in a form that can be used by fb_dyn (invoked by **str_cstr**). This function belongs to the multi-model sub-toolbox.

Synopsis.

```
CSTR = ktf_cstr(Nmin,Nmax,Dmin,Dmax,Matdeg)
```

Description. First, read page 208. Let K denote the feedback gain that is being computed. It is considered in transfer function matrix form.

$$K_{(i,j)}(s) = N_{(i,j)0} + \frac{N_{(i,j)1}s^{q-1} + \ldots + N_{(i,j)q}}{D_{(i,j)0}s^q + D_{(i,j)1}s^{q-1} + \ldots + D_{(i,j)q}}$$

The degrees (q) of the denominator of K(i,j) is given by Matdeg(i,j). The numerator coefficients can be written as follow (p number of measurements, m number of inputs)

$$\begin{bmatrix} \begin{bmatrix} N_{(1,1)0} & \cdots & N_{(1,p)0} \\ \vdots & & \vdots \\ N_{(m,1)0} & \cdots & N_{(m,p)0} \end{bmatrix} \\ \begin{bmatrix} N_{(1,1)1} & \cdots & N_{(1,p)1} \\ \vdots & & \vdots \\ N_{(m,1)1} & \cdots & N_{(m,p)1} \end{bmatrix} \\ \vdots \end{bmatrix}$$

with non relevant entries set to zero. **Nmin** and **Nmax** are matrix having the same structure, giving element-wise minimum and maximum values. Similar for **Dmin** and **Dmax** but the bounds are relative to the denominator (in this version of the toolbox **Dmin** = **Dmax**).

See also: **str_cstr, fb_dyn, kss_cstr**

Discussion. This function is useful for defining more specific constraints than those available *via* **str_cstr** and **add_cstr**. The input arguments **Nmin, Nmax, Dmin, Dmax** offer better visibility than a direct treatment of CSTR (see page 295).

3.2.35 Function: LSIM_MOD

Purpose. Simulation of the modal decomposition.

Synopsis.

```
[p_dom,12_norm] = lsim_mod(sys,u,t,x0,ixy[,npol[,sbplot1
                                   [,sbplot2[,sbplot3]]]])
```

Description. This function simulates the modal decomposition of one entry of the state vector or of one entry of the measurements vector. If it is invoked with output argument(s), the simulation is not displayed but the poles are sorted out, ordered with respect to their $\mathcal{L}_2$ contribution. Dominant modes can be identified in that way.

Caution. When this function is used for analysing dominant modes: sometimes modes that appear to be dominant cancel each others (see page 97), therefore, are not dominant.

Input arguments.

sys,u,t,x0   Same as for the function lsim *or alternatively* u = 'u1' (or 'u2',...) for an unit step demand at the first (or second,...) input. In this case t indicates the length of the simulation.

ixy   String like 'x1' (or 'x2',...), 'y1' (or 'y2',...) which indicates the signal (only one at a time) which is plotted. 'xi' stands for the ith state, 'yi' for the ith output.

npol   Is the number of modes which are plotted. The selection is made by considering the largest $\mathcal{L}_2$ norms during the simulation interval. If npol has 2 entries the modes from rank npol(1) to rank npol(2) are plotted.

sbplot1   Is a 1 by 3 integer matrix giving the sub-plot box for displaying the modal simulation result.

sbplot2   Similar for the legend.

sbplot3   Similar for the global simulation. The exact global simulation corresponds to the solid line. The sum of the npol dominant modes corresponds to the dashed line.

For example sbplot1=[1,2,1], sbplot2=[2,2,2], sbplot3=[2,2,4].

Output arguments.

p_dom   If there is an output argument the simulation is not plotted. p_dom is the vector of the npol dominant poles (ordering in decreasing dominance).

12_norm   Is the vector of the $\mathcal{L}_2$ contribution measures of the modes.

See also: `lsim, plot_res`

Examples. See pages 98, 133. This function is used for dominant poles identification on page 162. The following commands lines illustrate modal analysis (note that we close the feedback loop before using `lsim_mod`).

```
sys = rcamdata('lat',0,2,0);
Klat = fb_prop(sys,0,[-1+i;-.6;-1.7;-1.4;-1.2],'n',...
                                      [4 4 1 1 1]);
sysfb = feedback(sys,Klat,1);

lsim_mod(sysfb,'u1',10,0,'y1')
```
The resulting modal, decomposition is given in Figure 3.3.

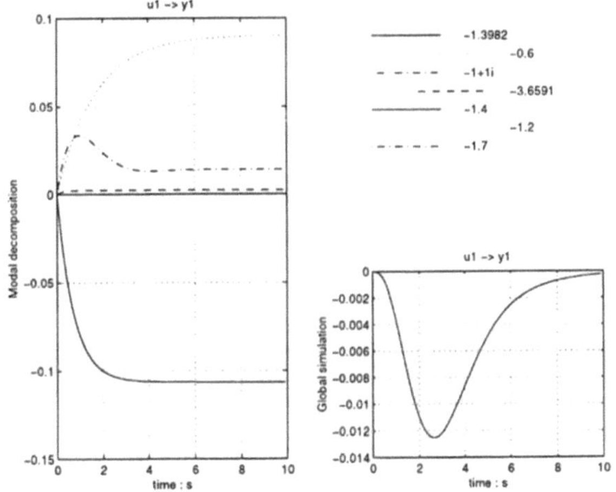

Figure 3.3. Function `lsim_mod`: A modal decomposition

Discussion. See §1.5.1, page 96.

3.2.36   Function: **OB_GENE**

Purpose. Observer design. Dual to `defin_vw`.

Synopsis.
```
[U,T,Pi] = ob_gene(sys,pol,key,def_pb[,bb,dd]);
```

Description. Observer design by juxtaposition of elementary observers *i.e.*, computation of the matrices U, T and Pi. These matrices are defined in the following figure.

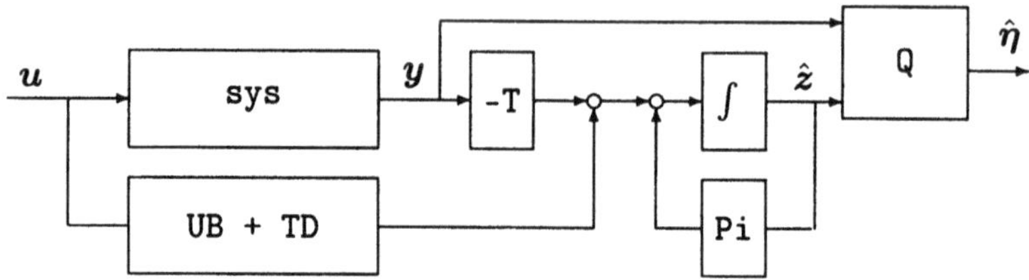

Each elementary observer design is considered as a dual right eigenvector assignment problem (see **help1 defin_vw**). The options are **'z'** and **'n'** for decoupling, **'p'**, **'v'** for projection or **'m'** for "dual minimum energy".

Input arguments.

sys   LTI system (see **ss.m**).

pol   Vector (of length q) of assigned eigenvalues. Do not repeat conjugate values.

key   Vector (of length q) defining options for elementary observer designs. key(i) is one of the strings **'z'**, **'i'**, **'p'**, **'n'** or **'v'**.

def_pb   Matrix having q columns. Non significant entries are set to zero. If key(i)

> == **'z'**: def_pb(:,i) contains the indices of the entries of the eigenvector corresponding to pol(i) that must be set to zero.
>
> == **'m'**: minimum energy assignment, def_pb(1,i) contains the open-loop poles that must be shifted to pol(i) with "minimum energy".
>
> == **'p'**: for projection of open-loop eigenvector. Three cases
>
> 1- one open-loop real (resp. nonreal) pole replaced by a closed-loop real (resp. nonreal) one: pol(i) is the assigned (closed-loop) eigenvalue, def_pb(1,i) is the concerned open-loop eigenvalue.
>
> 2- two open-loop real poles OL_pole_1 and OL_pole_2 replaced by one closed-loop nonreal pole CL_pole. Then, pol(i) = CL_pole and def_pb(1,i) = OL_pole_1 + j OL_pole_2, def_pb(2,i) = 1.
>
> 3- one open-loop nonreal OL_pole replaced by two closed-loop real CL_pole_1 and CL_pole_2. Then, pol(i) = CL_pole_1 + j CL_pole_2, def_pb(1,i) = OL_pole, def_pb(2,i) = 2.

== 'n': def_pb(:,i): The vectors ui (left eigenvector) and ti (output directions) corresponding to pol(i) will be such that ui*bb(:,def_pb(:,i)) + ti*dd(:,def_pb(:,i)) = 0. Default, bb = sys.b , dd = sys.d.

== 'v': def_pb(:,i) is n by 1 (n number of states), it is the transconjugate of the desired vector ui (assigned by least squares minimization).

Output arguments.

U,T,Pi   Matrices defining an observer ($z =$ U x is observed.)

See also: fb_prop, defin_vw, sob_proj, ob_ins, add_obs

Examples. See pages 121, 137 and 138.

Discussion. For the meaning of the matrices U, T, Pi, see §1.2.7, page 47. A more general treatment of observers can be found in [Magni, 1996, Magni and Mouyon, 1991, Magni and Mouyon, 1994].

3.2.37   Function: **OB_INS**

Purpose. Observer design with optimization.

Synopsis.
```
[U,T,Pi,crit] =
       ob_ins(sys,pol[,niter[,cc[,op_crit[,op_min]]]]);
```

Description. This function optimizes the condition number of a matrix [cc;U] (in which cc*x are the measurements to be used for feedback and U*x are the observed signals). It is also possible to minimize the feedback gain norm. The "optimization" is initialized using an heuristic approach which only permits to avoid very bad condition numbers. Then, standard optimization is applied. The observer structure is defined on page 219.

Input arguments.

sys   LTI system (see ss.m). The matrices sys.b and sys.d are ignored.

pol   Vector of observation dynamics. Do not repeat conjugate values.

niter   Number of iterations. More precisely, niter(1) is the number of random trials before optimization, niter(2) is the number of iterations of the heuristic algorithm dual to the one of §1.2.3, niter(3) is the number of iterations of standard optimization. If this argument is scalar, `niter` is internally replaced by `niter*[1 1 1]`. If this argument is omitted `niter` is considered by default as equal to the number of states of `sys`.

cc   Sub-matrix of `sys.c` used for optimization. It must correspond to the measurements that will be used for feedback implementation. Default `cc = sys.c`. Example: `cc = []` for Kalman filter structure (see page 50).

op_crit   = `'k'` for minimization of the "observer gain" norm, = `'c'` for optimization of the condition number (default), = `[coef_k coef_c]` (two real numbers) for a linear combination of both above criteria.

op_min   Vector for optimization options (see `optimset`).

Output arguments.

U,T,Pi Matrices defining an observer. (The signal `z = U x` is observed.)

crit Values of the criterion of Equation (1.45) at the various stages of the optimization (`crit(1)` = initial value, `crit(2)` = value after random trials, `crit(3)` = after heuristic optimization, `crit(4)` = optimal value).

See also: `ob_proj, sob_proj, sfb_ins`

Example. Design of an observer with Kalman filter structure. First is computed an 8-order observer:

```
sys = rss(8,3,3);
pol=[-10;-1+i;-5+5*i;-2;-2+2*i];
[U,T,Pi,crit] = ob_ins(sys,pol,100,[]);
```

Then the observer and the system are connected.

```
Q = [zeros(8,3) inv(U)];
sysobs = add_obs(sys,U,T,Q);
```

After connection, the output vector of the system (`sysobs`) is an estimate of the state vector (see page 51).

Discussion. For the meaning of the matrices U, T, Pi, see §1.2.7, page 47. The heuristic intermediate algorithm consists of maximizing the angles between the left eigenvectors and the rows of the matrix "*C*" (see page 34 for the dual problem). See also [Kaustky and Nichols, 1990, Kaustky et al., 1985, Moore and Klein, 1976].

3.2.38   Function: OBS2DFB

Purpose. Transforms an observer-based feedback into a standard dynamic feedback.

Synopsis.
```
[fb,ff] = obs2dfb(sys,U,T,Q,K[,Pi]);
```

Description. See the figure of page 219 for the meaning of the matrices U, T, Pi, Q. The closed-loop system `feedback(sys,fb,1)*ff` will be equivalent to the system of this figure in which $u = K\hat{\eta} + v$.

Input arguments.

sys   LTI system (see `ss.m`). It is the nominal system that was used for designing the observer.

U,T   Matrices characterizing an observer.

Q   Matrix mixing measurements and observer states. Notation: (n denotes the number of states. The state, output and observed vectors are denoted x, y, z).

- State observer with Kalman filter structure: U is n by n and Q = `[zeros(size(T)) inv(U)]`.

- Minimum order state observer: `[C;U]` is n by n and Q = `inv([C;U])`.

- To observe Lx, Q should be s.t. `Q*[C;U]` = L.

- To observe `[C;U]`x, set Q = 1.

K   Proportional feedback applied to Q$[y;\hat{z}]$, where $\hat{z}$ is the state vector of the observer (estimate of $z = U\,x$), see the figure of page 219.

Pi   This input argument is useful only if the observer order is larger than the number of states because in this case Pi is not unique.

Output arguments.

fb   Observer-based dynamic feedback (LTI system), see Equation (1.75).

ff   Feedforward such that the observer based design is equivalent to `feedback(sys,fb,1)*ff`, see Equation (1.76)

See also: `add_obs, dfb2obs`

Examples. See pages 121 and 123.

Discussion. For the meaning of the matrices U, T, Pi, see §1.2.7, page 47. For Q, see page 50. In fact, the matrix Q is the matrix $[Q_y \; Q_z]$ of

Figure 1.7. The computations performed by this function are detailed on page 56.

3.2.39   Function: PLOT_CON

Purpose. Computes and plots (bar-diagram) an input/output controllability measure.

Synopsis.
```
[pol,con] = plot_con(sys,iu,iy[,npol[,fsize[,color]]]])
```

Description. The input / output controllability measure is considered between the iuth input(s) and the iyth output(s) (note that iu and iy can be vectors). For computing the global controllability measure of a system with m inputs and p outputs set iu = [1:m] and iy = [1:p]. A measure of the standard controllability corresponds to iy = [] (no outputs) and of the standard observability corresponds to iu = [] (no inputs).

If this function is invoked with output arguments, the bar-diagram is not displayed.

Input arguments.

sys   LTI system (see **ss.m**).

iu,iy   Indices of input(s) and output(s) for computation of the input / output controllability measure. iu = [] for observability measure. iy = [] for input controllability measure.

npol   Number of most significant poles to be considered.

fsize   Size of the fonts, default **fsize** = 10.

color   String like 'g-' for green solid line....

Output arguments.

pol   Vector of the **npol** poles with largest controllability (in the decreasing order of magnitude).

con   Measure of the controllability.

See also: **lsim_mod, plot_res**

Example. controllability, observability and "input / output controllability" are respectively considered:

```
sys=demodata(2);
figure
plot_con(sys,[1:2],[]);
figure
plot_con(sys,[],[1:4]);
figure
plot_con(sys,[1:2],[1:4]);
```

The measure of standard controllability relative to the poles is illustrated in Figure 3.4, observability in Figure 3.5 and global input / output controllability in Figure 3.6.

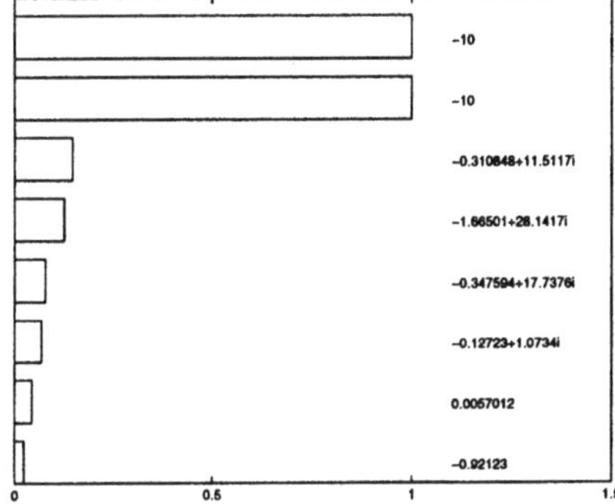

Figure 3.4. Function **plot_con**: Standard controllability.

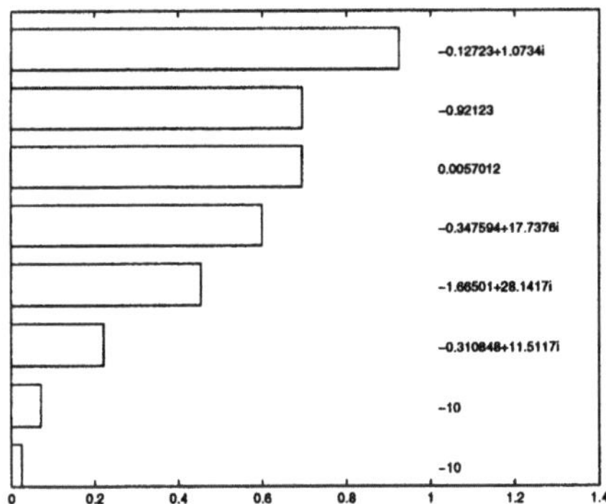

Figure 3.5. Function **plot_con**: Observability.

Concerning this specific example, it is worth noting that the poles with the highest input controllability have the worst observability. The input / output measure is more reliable.

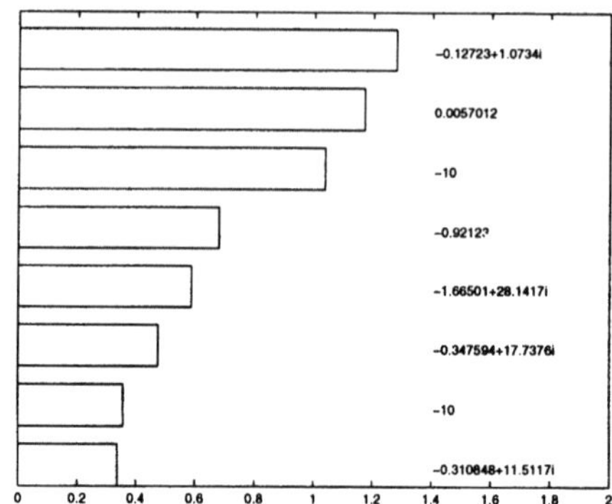

Figure 3.6. Function `plot_con`: Input / output controllability.

Discussion. See §1.5.2, page 99. In some cases, `plot_con` and `plot_res` lead to the same result.

3.2.40 Function: PLOT_RES

Purpose. Computes and plots (bar-diagram) residuals.

Synopsis.
```
[pol,res] = plot_res(sys,iu,iy[,time[,npol[,fsize[,color]]]])
```

Description. The residuals are considered between the iuth input(s) and the iyth output(s) (note that iu and iy can be vectors). If the input argument `time` is equal to 0, standard residuals are considered. If `time > 0`, residuals are replaced by the unit step response value at the given time (which can be `time = inf`). For computing the global residuals of a system with m inputs and p outputs set `iu = [1:m]` and `iy = [1:p]`.

If this function is invoked with output arguments, the bar-diagram is not displayed.

Input arguments.

sys   LTI system (see `ss.m`).

iu,iy   Indices of input(s) and output(s) for residual computation.

npol   Number of most significant poles to be considered.

`time` Scalar number. `time` = 0 for true residuals (impulse response), `time` = `inf` for residuals related to step response, 0 < `time` < `inf` for intermediate values of the step response.

`fsize` Size of the fonts, default `fsize` = 10.

`color` String like `'g-'` for green solid line....

Output arguments.

`pol` Vector of the `npol` poles with largest residuals (in decreasing order of magnitude).

`res` Values of the residuals.

See also: `lsim_mod, plot_con`

Example. Comparison of information obtained by using `lsim_mod` and `plot_res`: First, the true residuals (the 8 most significant ones) are computed:

```
randn('seed',0); rand('seed',0);
sys = rss(20,3,3);
figure
plot_res(sys,2,3,0,8);
```

The results are given in Figure 3.7.

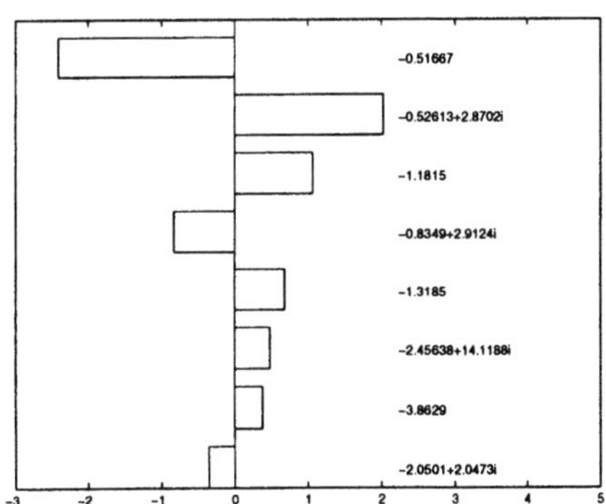

Figure 3.7. Function `plot_res`: Residuals.

Then, modal simulation is considered:

```
u = zeros(100,3);
t = 0.01*[0:1:99]';
x0 = zeros(20,1);
```

```
u(1,2)=100;
lsim_mod(sys,u,t,x0,'y3',8,[1,2,1],[1,2,2]);
```
The magnitude of the bars in Figure 3.7 corresponds to the initial values
of the modal simulation of Figure 3.8. Note that the ordering of poles is
not exactly the same using **plot_res** and **lsim_mod** because the measures
that are used are different.

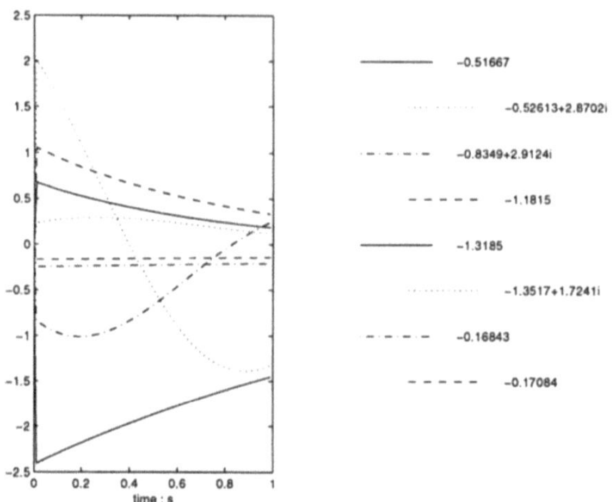

Figure 3.8.   Function **plot_res**: Modal simulation for comparison with residuals.

Now, let us consider a similar comparison concerning steady state values
of a step response.
```
figure
plot_res(sys,2,3,inf,8);
lsim_mod(sys,'u2',60,0,'y3',8,[1,2,1],[1,2,2]);
```
Here, the bars correspond to the final values of the simulation.

Discussion. See §1.5.1, page 96.

3.2.41   Function: PLOT_ZER

Purpose. 3-D plot of the minimum singular value of the system matrix.
Almost zeros of **sys** are the local minima (provided that these minima
are small enough).

Synopsis.
```
[gridx,gridy,Zm] = plot_zer(sys,gridx,gridy);
```

Input arguments.

sys   LTI system (see ss.m).

gridx   Gridding of the x-axis.

gridy   Gridding of the y-axis.

Output arguments.

gridx,gridy,Zm   See help mesh.

See also: azer

Example. Using the following commands:

```
sys = demodata(2);
[gridx,gridy,Zm] = plot_zer(sys,[-15:1.5:15],[0:1.5:30]);
mesh(gridx,gridy,Zm)
```

we obtain Figure 1.20 (page 102). This figure indicates the existence
of some zeros or almost zeros. The function **plot_zer** can be used for
initialization of the function **azer** (almost zero computation). To decide
if the results are true zeros (minimum equal to zero), relevant almost
zeros or irrelevant "almost zeros", the user must analyze the values of the
local minima.

3.2.42   Function: RCAMDATA

Purpose. Generates a bank of longitudinal and lateral linearized models
of a large carrier aircraft.

Synopsis.

```
[sys,b2,d2] = rcamdata(lonlat[,imodel[,itype[,delay]]])
```

Description. Notations: β is the angle of sideslip, p is the roll rate, r is
the yaw rate, ϕ is the roll angle, u_b, v_b and w_b are the three components
of the inertial velocity in the body-fixed reference frame, q is the pitch
rate, θ is the pitch angle, v_g is the ground speed. The architecture with
integrators is illustrated in Figure 5.1, page 292. International System
units are used.

The states, inputs and measurements are:

I - Lateral case (lonlat = 'lat').

- States
  itype = 0 $\rightarrow$ four states (p, r, ϕ, v_b)

`itype = 1` → plus two actuator states, plus four states for delay approximation (Padé filters)
`itype = 2` → plus two integrator states (on β and ϕ).

- Inputs
  `itype = 0 or 1` → two inputs for feedback (aileron and rudder)
  `itype = 2` → plus two inputs at integrator inputs (β_c, ϕ_c).

- Measurements
  `itype = 0 or 1` → four outputs for feedback (β, p, r, ϕ)
  `itype = 2` → plus two measurements at integrator outputs $(\int \beta, \int \phi)$.

II - Longitudinal case (`lonlat = 'lon'`)

- States
  `itype = 0` → four states $(q,\ \theta,\ u_b,\ w_b)$.
  `itype = 1` → plus two states corresponding to actuator and engine dynamics, plus four states for delay approximation (pade filters)
  `itype = 2` → plus two integrator states.

- Inputs `itype = 0 or 1` → two inputs for feedback (elevator and thrust)
  `itype = 2` → plus two inputs at integrator inputs (v_{zc}, v_{gc}).

- Measurements `itype = 0 or 1` → 4 outputs for feedback (v_z, v_g, q, n_z)
  `itype = 2` → plus two measurements at integrator outputs $(\int v_z, \int v_g)$.

Input arguments.

`lonlat` = `'lon'` or `'lat'` for longitudinal or lateral models.

`imodel` Model number from 0 (nominal) to 28. Default value: 0.

`itype` Equal to 0, 1 or 2. Default value: 2.

`delay` Delay at inputs (in seconds). Default value: 0.

Output arguments.

`sys` LTI representation of the system (see `ss.m`).

`b2,d2` Matrices for system inputs at intergartor inputs (used for simulation).

See also: `demodata`, `rcamstep`, `rcampole`

Discussion. Control design specifications are given on page 290. A table describing the available models is given on page 294. A non-linear precise description of the RCAM can be found in [Lambrechts et al., 1997].

3.2.43   Function: **RCAMPOLE**

Purpose. For demos. Plots the closed-loop poles of the models available using `rcamdata`.

Synopsis.
```
rcampole(lonlat,fb[,optn[,delay[,listm]]])
```

Input arguments.

`lonlat` = `'lon'` for the longitudinal case, = `'lat'` for the lateral case.

`fb`   Dynamic feedback gain (LTI system see `ss.m`) or proportional feedback.

`optn`   Is similar to the 3rd input argument of `rcamdata` (by default: `optn` = 2).

`delay`   Delay in seconds, `delay` = -1 means that delays are set to default values (by default: `delay` = -1).

`listm`   Vector of integers (model numbers) between 0 and 36. (Default: `listm` = [[0:1:28] [30:1:36]] as model 29 does not belong to the flight domain).

See also: `rcamstep, rcamdata, demodata`

Example. A feedback gain is computed using `fb_prop`. The corresponding pole map for all models (see `rcamdata`) is plotted using `rcampole`.
```
syslat = rcamdata('lat',0,2,0);
Klat = fb_prop(syslat,0,[-1+i;-.6;-1.7;-1.4;-1.2],...
                            'n',[4 4 1 1 1]);
rcampole('lat',Klat,2,0);
```
Figure 3.9 shows the multi-model pole assignment results.

Discussion. See page 290. The MATLAB symbols used for plotting the closed-loop poles are listed on page 294.

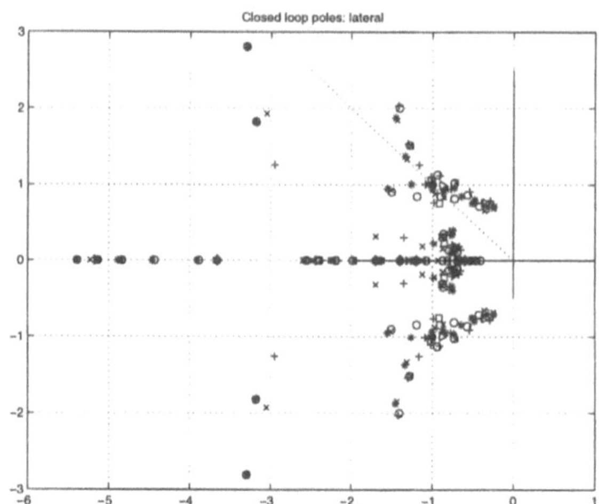

Figure 3.9. Function **rcampole**: Closed-loop poles of a bank of models

3.2.44   Function: RCAMSTEP

Purpose. For demos. Plots the closed-loop step responses of the models available using **rcamdata**.

Synopsis.

```
rcamstep(lonlat,fb[,ff[,optn[,delay[,listm]]]])
```

Input arguments.

lonlat = 'lon' for the longitudinal case, = 'lat' for the lateral case.

fb   Dynamic feedback gain (LTI system, see **ss.m**) or proportional feedback.

ff   Dynamic feedforward gain (LTI system, see **ss.m**) or constant gain (2 by 2 identity matrix by default).

optn   Similar to the 3rd input argument of **rcamdata** (default: optn = 2).

delay   Delay in seconds, **delay** < 0 means that delays are set to default values (default: **delay** = -1).

listm   Vector of integers (model numbers) between 0 and 36. (Default: listm = [[0:1:28] [30:1:36]] as model 29 does not belong to the flight domain).

See also: `rcampole, rcamdata, demodata`

Example. A feedback gain is computed using `fb_prop`. The corresponding step responses for all models (see `rcamdata`) is plotted using `rcamstep`.

```
syslat = rcamdata('lat',0,2,0);
Klat = fb_prop(syslat,0,[-1+i;-.6;-1.7;-1.4;-1.2],...
                                  'n',[4 4 1 1 1]);
rcamstep('lat',Klat,eye(2,2),2,0);
```

Figures 3.10 shows the simulation results.

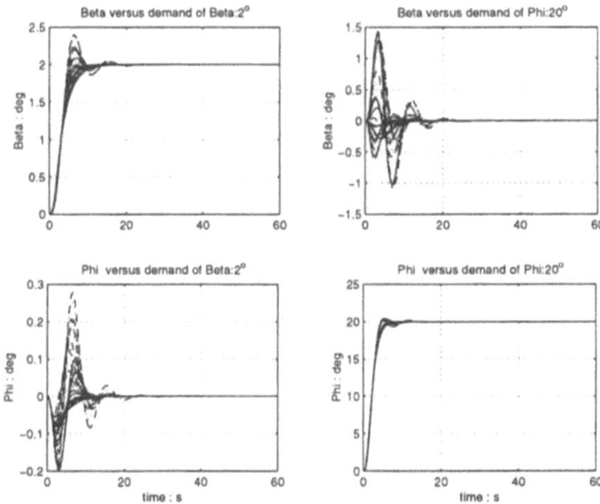

Figure 3.10. Function **rcamstep**: Step responses of a bank of models

Discussion. See page 290. The MATLAB symbols used for plotting the step responses are listed on page 294.

3.2.45   Function: SFB_INS

Purpose. State feedback design with optimization.

Synopsis.
```
[k,V,W,crit] =
    sfb_ins(sys,k0[,pol[,niter[,opt[,op_crit[,op_min]]]]]);
[V,W] =
    sfb_ins(sys,k0[,pol[,niter[,opt[,op_crit[,op_min]]]]]);
```

Description. This functions assigns eigenstructure with maximal angle between the right eigenvectors ($\Leftrightarrow$ insensitivity of eigenvalues). It is also possible to minimize the feedback gain norm. Exact eigenvalue assignment is insured in the state feedback case. But there is *no guarantee of stability in the output feedback case* (option 'o') because in this case, some poles are not controlled. The "optimization" is initialized using an heuristic approach (page 34) which only permits us to avoid very high sensitivity. Then, standard optimization is applied.

Caution. Exact pole assignment is relevant for low dimensional systems. In case of weakly controllable poles (see plot_con), use partial pole assignment.

Input arguments.

sys   LTI system (see ss.m).

k0   Initial gain (k0=0 means that k0 is the zero matrix with appropriate size). Usually, this argument is used when it is intended to pursue an optimization by setting k0 = k. k0 is a state feedback gain if opt = 's' or an output feedback gain if opt = 'o'.

pol   Vector of assigned eigenvalues, conjugate values are not repeated. If k0 is not equal to 0, the entries of pol must be close enough to the closed-loop eigenvalues. If pol is omitted, it is considered as the vector of closed-loop eigenvalues.

niter   Number of iterations. More precisely, niter(1) is the number of random trials before optimization, niter(2) is the number of iterations of the heuristic algorithm of §1.2.3, niter(3) is the number of iterations of standard optimization. If this argument is scalar, niter is internally replaced by niter*[1 1 1]. If this argument is omitted niter is considered by default as equal to the number of states of sys.

op_crit   = 'k' for minimization of the gain norm, = 'c' for optimization of the condition number (default), = [coef_k coef_c] (two real numbers) for a linear combination of both above criteria.

opt   The option opt = 's' (default) means "state feedback" (in this case, the matrices sys.c and sys.d are ignored). The option opt = 'o' means output feedback.

op_min   Vector for optimization options (see optimset).

Output arguments.

k,V,W Computed feedback gain, assigned eigenvectors and input directions.

crit Values of the criterion of Equation (1.45) at the various stages of the optimization (crit(1) = initial value, crit(2) = value after random trials, crit(3) = after heuristic optimization, crit(4) = optimal value).

See also: fb_prop, sfb_proj, ob_ins, dfb_ins

Example 1. Consider a random system with as many measurements as states.

```
sys = rss(8,8,3);
pol=[-1;-2+2*i;-4+4*i;-4;-1+i];
k=sfb_ins(sys,0,pol,20,'o');
```

For additional iterations, in order to insure continuity, the result k is "re-injected" as an input argument.

```
k=sfb_ins(sys,k,pol,40,'o');
[eig(sys) eig_fb(sys,k)]
```

Example 2. Two measurements are missing for state feedback, so, a two-dimensional observer is designed.

```
sys = rss(8,6,3);
polo = [-3+3*i];
[U,T,Pi] = ob_ins(sys,polo,20);
sysobs = add_obs(sys,U,T,1);
```

then insensitivity is treated (by output feedback). In this case, stability is insured by the separation principle.

```
polc=[-1;-2+2*i;-4+4*i;-4;-1+i];
[ko,V,W,crit]=sfb_ins(sysobs,0,polc,200,'o');
eig_fb(sysobs,ko)
```

Discussion. The algorithm used for initialization is heuristic, it consists of maximizing the angles between the right eigenvectors, see pages 34, 122. See also [Moore and Klein, 1976, Kaustky and Nichols, 1990, Kaustky et al., 1985]. The use of this heuristic initialization has shown to be very efficient for avoiding convergence towards bad local minima.

3.2.46　Function: SFB_PROJ

Purpose. Computes a state feedback by projection of the open-loop eigenstructure.

Synopsis.

```
[k,V,W,pol] = sfb_proj(sys,Tr,Damp[,key[,C_ratio]]);
```

Description. The closed-loop poles are automatically chosen from the design parameters which are the minimum damping ratio (`Damp`) and the settling time (`Tr`). The right eigenvectors corresponding to open-loop eigenvalues satisfying the settling time and damping ratio requirements are re-assigned as they are. For other open-loop poles that need to be shifted, assignment of right eigenvectors is made by projection of the open-loop ones.

Caution. Do not use this function if there are repeated open-loop eigenvalues.

Input arguments.

sys　LTI system (see `ss.m`). The matrices `sys.c` and `sys.d` are ignored.

Tr　Settling time.

Damp　Minimum damping ratio for closed-loop eigenvalues ($0 < $ `Damp` $ < 1$).

key　Is a string equal to `'p'` for orthogonal projection of all eigenvectors, or equal to `'m'` for minimum energy assignment (default: `key` = `'m'`).

C_ratio　If the controllability degree of some eigenvalue is less than the maximum controllability degree divided by `C_ratio`, this eigenvalue is not shifted. Default value `C_ratio` = 100.

Output arguments.

k　State feedback.

V,W,pol　Assigned eigenstructure.

See also: `fb_prop, sob_proj, sfb_ins`

Example. Design of two state feedback gains such that the damping ratio becomes larger than 0.707 and the settling time less than 15 seconds. First design by open-loop projection, second by minimum energy assignment.

```
randn('seed',0); rand('seed',0);
sys = rss(6,6,3);

Tr = 15; Damp = 0.707;
kp = sfb_proj(sys,Tr,Damp,'p');
km = sfb_proj(sys,Tr,Damp,'m');
norm(km)
norm(kp)
```

The norm of the gain obtained with option 'm' is usually (not always) smaller than with option 'p'. For this example we obtain `norm(km)` = `5.79` and `norm(kp)` = `12.8`. Other example on page 119.

Discussion. See §2.2.1, page 118.

3.2.47 Function: SOB_PROJ

Purpose. Computes a full state observer with Kalman filter structure (see page 50) by projection of the open-loop left eigenstructure.

Synopsis.
```
[g,U,T,Pi,Q] = sob_proj(sys,Tr,Damp,key,O_ratio);
```

Description. The observer poles are automatically chosen form the design parameters which are the minimum damping ratio (`Damp`) and the settling time (`Tr`). The left eigenvectors corresponding to open-loop eigenvalues satisfying the settling time and damping ratio requirements are considered for elementary observations (see Lemma 1.2.5, page 47). For other open-loop poles that need to be shifted, elementary observations correspond to the projection of open-loop left eigenvectors.

Caution. Do not use this function if there are repeated open-loop eigenvalues.

Input arguments.

sys LTI system (see `ss.m`). The matrices `sys.b` and `sys.d` are ignored.

Tr Settling time.

Damp Minimum damping ratio for closed-loop eigenvalues $(0 < \text{Damp} < 1)$.

key Is a string equal to 'p' for orthogonal projection of eigenvectors, or equal to 'm' for dual minimum energy assignment.

O_ratio   If the controllability degree of some eigenvalue is less than the
maximum controllability degree divided by O_ratio, this eigenvalue
is not shifted. Default value O_ratio = 100.

Output arguments.

g   Observer gain (Kalman filter structure).

U,T,Pi,Q   Observer matrices (structure of page 219).

See also: ob_gene, add_obs, sfb_proj

Example. Design of a full state observer with observation dynamics
corresponding to damping ratio larger than 0.6 and settling time less
than 20 seconds.

```
randn('seed',0); rand('seed',0);
sys = rss(6,3,2);
```

```
[g,U,T,Pi,Q] = sob_proj(sys,20,0.6,'m');
```
This observer is connected to the system
```
sysobs = add_obs(sys,U,T,Q);
```
A state feedback is computed, it can be implemented as an output feedback of the system **sysobs**. For example:
```
k=sfb_proj(sys,20,0.6,'m');
```
From the separation principle we must have **eig_fb(sysobs,k)** equal to
the union of **eig(sys.a+sys.b*k)** and **eig(sys.a+g*sys.c)** (right hand
side vector below).

```
-3.2806 + 4.3741i   -3.2406 + 4.3208i
-3.2806 - 4.3741i   -3.2406 - 4.3208i
-3.2406 + 4.3208i   -3.2569 + 3.2855i
-3.2406 - 4.3208i   -3.2569 - 3.2855i
-3.2569 + 3.2855i   -2.4500
-3.2569 - 3.2855i   -1.3665
-3.2155 + 3.2437i   -3.2806 + 4.3741i
-3.2155 - 3.2437i   -3.2806 - 4.3741i
-2.4512             -3.2155 + 3.2437i
-2.4500             -3.2155 - 3.2437i
-1.3611             -2.4512
-1.3665             -1.3611
```

Discussion. See page 121.

3.2.48   Function: SORT_EV

Purpose. Sorts out the entries of a vector which satisfy a set of logical expressions of the form "real part (or imaginary part or damping ratio) more than (or less than or equal to)".

Synopsis.
```
[ll,ind] = sort_ev(lambda,ri1,ie1,b1[,ao1,ri2,ie2,b2
                       [,ao2,ri3,ie3,b3[,ao3,ri4,ie4,b4]]]);
[ll,ind] = sort_ev(lambda,expr);
```

Input arguments.

`lambda`   Vector of the values that must be treated.

`ri1,ri2,ri3,ri4`   Strings like 'r' (real part), 'i' (imaginary part), 'd' (damping ratio).

`ie1,ie2,ie3,ie4`   Strings like '<', '>', '>=', '<=', '=='

`b1,b2,b3,b4`   Scalars.

`ao1,ao2,ao3`   Strings like '&', '|'.

`expr`   Concatenated above input arguments, but in this case, b1, b2... must be numerical values. For example:
```
b1 = 0; sort_ev(lambda,'i','>',b1);
```
is equivalent to
```
sort_ev(lambda,'i>0');
```
but `sort_ev(lambda,'i>b1');` is not valid (because the numerical value of `b1` is not defined in the function `sort_ev`).

Output argument.

`ll`   Result.

`ind`   Such that $ll = lambda(ind)$.

See also: `vicin_ev`, `clean_ev`, `choi_ev`, `contr_ev`

Examples. See pages 89, 146 and 156.

3.2.49   Function: STA2DYN

Purpose. Transforms a proportional feedback gain relative to a system that was beforehand extended using **add_dyn**, into an equivalent dynamic gain relative to the non-extended original system.

Synopsis.

```
fbdyn = sta2dyn(kprop,dyn);
```

Description. See add_dyn page 175.

Input arguments.

kprop   Proportional feedback gain relative to an extended system.

dyn   Vector that was used as input argument when the system was extended using add_dyn.

Output argument.

fbdyn   LTI system (see ss.m): dynamic feedback gain equivalent to kprop.

See also: add_dyn, dyn2sta

Examples. See pages 46, 145 and 205.

Discussion. See page 46.

3.2.50   Function: STR_CSTR

Purpose. Defines the structure of a dynamic feedback (see fb_dyn).

Synopsis.

```
CSTR = str_cstr(m,p,pol[,dif_deg[,num_bds]])
```

Description. First, read page 208. Let us denote m and p the numbers of inputs and outputs of the considered system, $K(s)$ is m by p. The gain structure is for $i = 1, ..., m$ and $j = 1, ..., p$

$$K_{(i,j)}(s) = N_{(i,j)0} + \frac{N_{(i,j)1}s^{q-1} + ... + N_{(i,j)q}}{D_{(i,j)0}s^q + D_{(i,j)1}s^{q-1} + ... + D_{(i,j)q}}$$

This function permits the designer to choose:

- the degrees of the denominators and the coefficients $D_{(i,j)k}$ by listing the desired corresponding poles in pol.

- the degrees of the numerators and bounds relative to the numerator coefficients $N_{(i,j)k}$.

In the case in which all denominators are equal and numerator coefficients are not bounded, only three input arguments are required. If in addition, the difference between the numerator and denominator degrees is non-zero, the 4th optional argument **dif_deg** must be used. Numerator bounds can be defined, entry per entry, using the 5th input arguments **num_bds**.

The constraints computed by this function are written in CSTR in a form that can be treated by **fb_dyn**. This function belongs to the multi-model sub-toolbox. It must be used before all other functions of this sub-toolbox in order to initialize the matrix CSTR. It is recommended to check the gain structure using **str_view(CSTR)**.

Input arguments.

m,p Numbers of inputs and outputs of the system.

pol Defines the poles of the denominators. The conjugate values are not repeated. There are four possibilities:

1- all the gain entries have the same denominator $\rightarrow$ **pol** is a vector *e.g.*, **pol = [-1 -2+2i]** (or **pol = {[-1 -2+2i]}**)

2- same denominator for all entries in each row (filters at system inputs) $\rightarrow$ **pol** is a (m by 1) cell *e.g.*, **m=2, pol = {[-1 -2+2i];[-3+3i]}**

3- same denominator for all entries in each column (filters at system outputs) $\rightarrow$ **pol** is a (1 by p) cell *e.g.*, **p= 2, pol = {[-1 -2+2i] [-3+3i]}**

4- all denominators are distinct $\rightarrow$ **pol** is a (m by p) cell *e.g.*, $p = 2$, $m = 2$, **pol = {[-2] [-3];[-4 -5] [-1+i]}**

dif_deg Defines the difference between numerator and denominator degrees. Same four possibilities as above, respectively: **dif_deg** is a (1 by 1), (m by 1), (1 by p) or a (m by p) integer matrix.

num_bds Defines bounds on numerator coefficients. Polynomial coefficients are ordered in descending power: *e.g.*, **num_bds = { [1 2] [0 -2 -inf] [0 2 +inf]; [2 3] [1 -inf] [2 +inf]}** means that:

- for gain entry **(1,2)** (denominator degree = 2) $N_{(1,2)0} = 0$, $-2 < N_{(1,2)1} < 2$, $N_{(1,2)2}$ not constrained.

- for gain entry **(2,3)** (denominator degree = 1) $2 < N_{(2,3)0} < 3$, $N_{(2,3)1}$ not constrained. Note that the length of the vectors defining the numerator coefficient bounds is equal to the degree of the denominator + 1 (regardless of **dif_deg**).

See also: str_view, fb_dyn, fb_view, add_cstr (,ktf_cstr)

Examples. See pages 74, 75, 143 , 155, 160 and 162.

Discussion. The controller structure is detailed in Equation (1.85), page 65 (note that in the software, the gain entries might have different degrees). In the examples referred above, the gain structure that is used is elementary. See [Le Gorrec et al., 1998a] for an example of a more complex structure. The possibilities offered using **str_cstr** might not be sufficient, in this case the more general function **ktf_cstr** can be considered.

3.2.51   Function: **STR_VIEW**

Purpose. Displays to the screen the gain structure defined by **str_cstr**.

Synopsis.

```
str_view(CSTR[,entries])
```

Input arguments.

CSTR Output argument of **str_cstr**.

entries q by 2 matrix, each row defines one of the q gain entries that is to be displayed. Default: all entries are displayed.

See also: str_cstr, fb_dyn, fb_view

Example. The desired gain structure is

$$\left[\begin{array}{cc} k_1 & \frac{k_2}{s^2+4s+8} \\ 0 & k_3 + \frac{k_4}{s+3} \end{array} \right]$$

This structure is transformed to constraints by **str_cstr**.

```
pol = {[],[-2+2*i];[],[-3]};
dif_deg = [0 2;1 0];

CSTR = str_cstr(2,2,pol,dif_deg);
```

The constraints can be checked using **str_view**.

```
str_view(CSTR)
```

The message displayed to the screen by **str_cstr** indicates, entry per entry, the poles and the bounds of all numerator ceofficients.

3.2.52   Function: **VICIN_EV**

Purpose. Sorts out the entries of a vector close enough to the entries of an other reference vector.

Synopsis.

```
[lambda_in,lambda_out] = vicin_ev(lambda,lambda0[,radius]);
```

Input arguments.

lambda   Vector of the values that are to be treated.

lambda0,radius   Vectors having the same length, defining circles. The kth circle is defined by lambda0(k) (center) and by radius(k) (radius). The function vicin_ev finds the entries of lambda belonging to these circles. (Default: radius(k) = 10 percent of $|$lambda0(k)$|$).

Output argument.

lambda_in Entries of lambda (imaginary part $\geq$ 0) belonging to the circles defined by lambda0 and radius.

lambda_out complement of lambda_in (imaginary part $\geq$ 0) in lambda.

See also: sort_ev, clean_ev, choi_ev, contr_ev

Chapter 4

APPENDIX 1: PROOFS OF THE RESULTS STATED IN THE FIRST CHAPTER

Contents.

- Technical results for multi-model eigenvector assignment by dynamic feedback ....page 248

- First order perturbation of eigenstructure and multi-model phase control ....page 255

- Minimum energy assignment ....page 262

- Technical results for pole assignment by output feedback ..page 265

- Introduction to classical non-interactive control design and adaptation to eigenstructure assignment ....page 268

- Dynamic feedforward design generalizing SISO pole/zero cancellation to the MIMO case ....page 274

- Observers and observer-based feedback design ....page 278

It is assumed in this book that, any time non-real eigenvalues and eigenvectors are assigned, the corresponding complex conjugate eigenvalues and eigenvectors are also assigned. It is also assumed that assigned eigenvalues are never repeated.

A multivariable linear system is considered (see page 13 for more detailed notations):

$$\dot{x} = Ax + Bu$$
$$y = Cx + Du \tag{4.1}$$

in which $x \in \mathbb{R}^n$, $u \in \mathbb{R}^m$ and $y \in \mathbb{R}^p$. The transfer function matrix is denoted $S(s)$:

$$S(s) = C(sI - A)^{-1}B + D \tag{4.2}$$

Notations are often simplified considering $D = 0$. For this reason, the following straightforward result is often used in this appendix. Assuming that K is such that $(I - DK)$ is non-singular, both following feedback loops are equivalent:

- $\dot{x} = Ax + Bu$, $y = Cx + Du$, $u = Ky$

- $\dot{x} = Ax + Bu$, $y = Cx$, $u = K(I - DK)^{-1}y$

The artificial feedback $K(I - DK)^{-1}$ is denoted $\tilde{K}$. Note that we can compute K from $\tilde{K}$ as follows:

$$K = (I + \tilde{K}D)^{-1}\tilde{K} \tag{4.3}$$

Several matrix identities will be used:

- $(I - KD)^{-1}K = K(I - DK)^{-1}$

- $(I - KD)^{-1}KD = KD(I - KD)^{-1}$

- $(I - KD)^{-1} = I + K(I - DK)^{-1}D$

The Popov, Belevitch and Hautus controllability criterion is used several times in this appendix. This criterion states that an open-loop eigenvalue λ_i of a pair (A, B) is non-controllable if and only if there exists a nonzero vector u_i (left eigenvector) such that

$$u_i \begin{bmatrix} A - \lambda_i I & B \end{bmatrix} = 0$$

see [MacFarlane and Karcanias, 1976]) for example. The dual counterpart is also used: an open-loop eigenvalue λ_i of a pair (A, C) is non-observable if and only if there exists a nonzero vector v_i (right eigenvector) such that

$$\begin{bmatrix} A - \lambda_i I \\ C \end{bmatrix} v_i = 0$$

4.1   Results relative to eigenvector assignment

This section is devoted to the computation of feedback gains for eigenvalue and eigenvector assignment. First, is considered the case where the gain is a simple matrix, then, the transfer matrix case is treated. Finally, the Linear Quadratic Programming problem arising in the transfer function matrix case is written in an explicit form.

4.1.1   Proof of Lemmas 1.1.1 and 1.1.2

The first lemma (page 15) is recalled.

LEMMA 4.1.1 *A triple (v_i, w_i, λ_i) can be assigned by proportional output feedback if and only if it satisfies*

$$\begin{bmatrix} A - \lambda_i I & B \end{bmatrix} \begin{bmatrix} v_i \\ w_i \end{bmatrix} = 0 \qquad (4.4)$$

This triple is assigned in closed-loop by any gain K satisfying the linear equation

$$K(Cv_i + Dw_i) = w_i \qquad (4.5)$$

Proof. From (4.4)

$$Av_i + Bw_i = \lambda_i v_i$$

and from (4.5)

$$w_i = (I - KD)^{-1} KCv_i$$

Combining both equations

$$(A + B(I - KD)^{-1} KC)v_i = \lambda_i v_i$$

which proves the "if part" of the lemma. For the "only if part", consider the last equation written as

$$\begin{bmatrix} A - \lambda_i I & B \end{bmatrix} \begin{bmatrix} v_i \\ (I - KD)^{-1} Cv_i \end{bmatrix} = 0$$

defining $w_i = (I - KD)^{-1} Cv_i \Leftrightarrow K(Cv_i + Dw_i) = w_i$ ends the proof of the lemma. ∎

Lemma 1.1.2 is the exact dual counterpart of Lemma 1.1.1, so, it does not need to be proved.

When the feedback gain is computed as above, it is required for the existence of a solution that the matrices $C[v_1 \ldots v_q]$ and $(C[v_1 \ldots v_q] + D[w_1 \ldots w_q])$ are of maximum column rank. Some generic properties are known for that purpose, it can be shown that for almost any choice of the vectors v_i satisfying (4.4) the expected rank properties are satisfied. However, these results are of secondary importance for applications and are very technical, so, will be ignored here (see for example [Kimura, 1975] or [Champetier and Magni, 1991]).

4.1.2 Proof of Lemma 1.2.4

Before stating Lemma 1.2.4 (*i.e.*, Lemma 4.1.2, below), the dynamic extension of page 45 is recalled.

$$\begin{bmatrix} \dot{x} \\ \dot{x}_c \end{bmatrix} = \begin{bmatrix} A & 0 \\ 0 & \Omega \end{bmatrix} \begin{bmatrix} x \\ x_c \end{bmatrix} + \begin{bmatrix} B & 0 \\ 0 & I_{n_c} \end{bmatrix} \begin{bmatrix} u \\ u_c \end{bmatrix}$$

$$\begin{bmatrix} y \\ y_c \end{bmatrix} = \begin{bmatrix} C & 0 \\ 0 & I_{n_c} \end{bmatrix} \begin{bmatrix} x \\ x_c \end{bmatrix} + \begin{bmatrix} D & 0 \\ 0 & 0 \end{bmatrix} \begin{bmatrix} u \\ u_c \end{bmatrix} \qquad (4.6)$$

LEMMA 4.1.2 *The following feedback gains are equivalent:*

- $G(s)$ *applied to System (4.1), where*

$$G(s) = K_{11} + K_{12}(sI - K_{22} - \Omega)^{-1}K_{21} \qquad (4.7)$$

- $\overline{K}$ *applied to System (4.6), where*

$$\overline{K} = \begin{bmatrix} K_{11} & K_{12} \\ K_{21} & K_{22} \end{bmatrix} \qquad (4.8)$$

Proof. Applying the feedback (4.8) to (4.6) consists of closing the loop as follows

$$\begin{bmatrix} u \\ u_c \end{bmatrix} = \begin{bmatrix} K_{11} & K_{12} \\ K_{21} & K_{22} \end{bmatrix} \begin{bmatrix} y \\ y_c \end{bmatrix} \qquad (4.9)$$

This equation is equivalent to

$$u = K_{11}y + K_{12}y_c \text{ and } u_c = K_{21}y + K_{22}y_c \qquad (4.10)$$

From (4.6):

$$y_c = x_c \text{ and } (sI - \Omega)x_c = u_c = K_{21}y + K_{22}y_c$$

The last equation can be written

$$x_c = (sI - K_{22} - \Omega)^{-1}K_{21}y \qquad (4.11)$$

Combining (4.10) and (4.11):

$$u = K_{11}y + K_{12}(sI - K_{22} - \Omega)^{-1}K_{21}y$$

which is $u = G(s)y_c$ i.e., $G(s)$ applied to System (4.1) as stated in the lemma. The converse is proved backward by considering the last equation into which is substituted x_c defined by (4.11) and so on. ∎

4.1.3   Proof of Lemma 1.3.1

In the case where the gain is a transfer function matrix instead of a real matrix, Lemma 4.1.3 must be used instead of Lemma 4.1.1. The difference between both results reduces to the fact that "K" is now a transfer function matrix $G(s)$, so, it is $G(\lambda_i)$ (a complex constant matrix) that must be considered instead of K. Lemma 1.3.1 is recalled:

LEMMA 4.1.3   *A triple* (v_i, w_i, λ_i) *which satisfies (4.4) can be assigned in closed-loop by any dynamic gain* $G(s)$ *satisfying the linear equation*

$$G(\lambda_i)(Cv_i + Dw_i) = w_i \tag{4.12}$$

Proof. This lemma is proved by combining Lemmas 4.1.1 and 4.1.2. Assume that a triple (λ_i, v_i, w_i) satisfies (4.4). It follows that for $w_{ci} = \lambda_i v_{ci}$, so

$$\begin{bmatrix} A - \lambda_i & 0 & B & 0 \\ 0 & -\lambda_i I & 0 & I \end{bmatrix} \begin{bmatrix} v_i \\ v_{ci} \\ w_i \\ w_{ci} \end{bmatrix} = 0 \tag{4.13}$$

Therefore, in view of Lemma 4.1.1, the feedback gain that assigns the following triple

$$\left(\lambda_i, \begin{bmatrix} v_i \\ v_{ci} \end{bmatrix}, \begin{bmatrix} w_i \\ w_{ci} \end{bmatrix} \right)$$

for the extended system (4.6), satisfies

$$\begin{bmatrix} K_{11} & K_{12} \\ K_{21} & K_{22} \end{bmatrix} \left(\begin{bmatrix} C & 0 \\ 0 & I \end{bmatrix} \begin{bmatrix} v_i \\ v_{ci} \end{bmatrix} + \right.$$

$$\left. \begin{bmatrix} D & 0 \\ 0 & 0 \end{bmatrix} \begin{bmatrix} w_i \\ w_{ci} \end{bmatrix} \right) = \begin{bmatrix} w_i \\ w_{ci} \end{bmatrix} \tag{4.14}$$

Equations (4.13) and (4.14) are equivalent to the following set of four equations:

$$\begin{bmatrix} A - \lambda_i I & B \end{bmatrix} \begin{bmatrix} v_i \\ w_i \end{bmatrix} = 0$$

$$\lambda_i v_{ci} = w_{ci}$$

$$K_{11} C v_i + K_{12} v_{ci} + K_{11} D w_i = w_i$$

$$K_{21} C v_i + K_{22} v_{ci} + K_{21} D w_i = w_{ci} = \lambda_i v_{ci}$$

The fourth equation can be written as $v_{ci} = (\lambda_i I - K_{22})^{-1} K_{21}(C v_i + D w_i)$, after substitution of v_{ci} the third equation becomes:

$$K_{11}(C v_i + D w_i) + K_{12}(\lambda_i I - K_{22})^{-1} K_{21}(C v_i + D w_i) = w_i$$

or

$$(K_{11} + K_{12}(\lambda_i I - K_{22})^{-1} K_{21})(C v_i + D w_i) = w_i$$

or

$$G(\lambda_i)(C v_i + D w_i) = w_i$$

The proof of the converse is similar: Defining both vectors v_{ci} and w_{ci} as above, from the last equation together with (4.4) it is straightforward to recover (4.13) and (4.14). ∎

4.1.4 Linear Quadratic Programming

In §4.1.3, the main result is proved using a state space approach. For numerical implementation, the state space approach is not very interesting because it leads to nonlinear equations. The approach used in this toolbox consider the gain matrix as a transfer function matrix having the form given in Equation (1.85), page 65 (with possibly distinct denominator degrees).

On page 66 it is just mentioned that solving Equation (4.12) with $G(s)$ as in (1.85) can be viewed as a Linear Quadratic Programming Problem (LQP Problem). The problem of obtaining an explicit LQP form is considered in Lemma 4.1.4 stated below.

Unfortunately, the general form of the transfer matrix adopted in the toolbox (function fb_dyn) induces cumbersome notations. Therefore, it is proposed to adopt a simpler form for the justification. So, the transfer matrix form of Equation (1.85) is replaced by the one of Equation (4.15).

$$G(s) = \begin{bmatrix} \dfrac{b_{011}s^q + b_{111}s^{q-1} + \ldots + b_{q11}}{s^q + a_{111}s^{q-1} + \ldots + a_{q11}} & \cdots & \dfrac{b_{01p}s^q + b_{11p}s^{q-1} + \ldots + b_{q1p}}{s^q + a_{11p}s^{q-1} + \ldots + a_{q1p}} \\ \cdots & \cdots & \cdots \\ \cdots & \cdots & \cdots \\ \dfrac{b_{0m1}s^q + b_{1m1}s^{q-1} + \ldots + b_{qm1}}{s^q + a_{1m1}s^{q-1} + \ldots + a_{qm1}} & \cdots & \dfrac{b_{0mp}s^q + b_{1mp}s^{q-1} + \ldots + b_{qmp}}{s^q + a_{1mp}s^{q-1} + \ldots + a_{qmp}} \end{bmatrix} \tag{4.15}$$

The quadratic criterion that is considered is defined in Equation (1.86) page 66.

$$J = \sum_{i}^{r} \|G_{\text{ref}}(j\omega_i) - G(j\omega_i)\|_F^2 \tag{4.16}$$

The assignment constraints consist of a set of equations similar to (4.12). These constraints are nonlinear but, by fixing the denominators, the dependency on free parameters (coefficients b_{ijk}) becomes linear[1]. In the sequel, it is assumed that denominator coefficients a_{ijk} are fixed. The following notation is necessary to exhibit the vector of free parameters. The row vector ξ_k contains the free parameters on the kth row of (4.15)

$$\xi_k = [b_{0k1} \dots b_{qk1} \dots \dots \dots b_{0kp} \dots b_{qkp}] \tag{4.17}$$

the vector of all free parameters is denoted ξ

$$\xi = [\xi_1 \dots \xi_m] \tag{4.18}$$

and

$$X_k(\lambda) = \text{Diag}\left\{ \frac{1}{a_{k1}(\lambda)} \begin{bmatrix} \lambda^q \\ \lambda^{q-1} \\ \vdots \\ 1 \end{bmatrix}, \dots, \frac{1}{a_{kp}(\lambda)} \begin{bmatrix} \lambda^q \\ \lambda^{q-1} \\ \vdots \\ 1 \end{bmatrix} \right\} \tag{4.19}$$

where $a_{kj}(\lambda)$ denotes the denominator of the kjth entry of $K(s)$ evaluated at $s = \lambda$. It follows that:

$$K(s) = \sum_{k=1}^{m} e_k \xi_k X_k(s) \tag{4.20}$$

where e_k is the kth vector of the canonical basis of $\mathbb{R}^m$ (*i.e.*, a m-dimensional column vector with all entries equal to zero except the kth one that is equal to one).

LEMMA 4.1.4 *The problem of finding $G(s)$ such that Equation (4.12) is satisfied and such that the criterion of (4.16) is minimum is a LQ-programming problem $\xi A_1 = b_1$, $J = \xi H \xi^T + 2\xi c$ in which:*

$$A_1 = \text{Diag}\{X_1(\lambda_i)(Cv_i + Dw_i) \ \dots \ X_m(\lambda_i)(Cv_i + Dw_i)\}$$

$$H = \text{Diag}\left\{ \sum_{i=1,\dots,r} X_1(j\omega_i)X_1(j\omega_i)^* \ \dots \ \sum_{i=1,\dots,r} X_m(j\omega_i)X_m(j\omega_i)^* \right\}$$

$$b_1 = \begin{bmatrix} e_1^T w_1 \\ \vdots \\ e_m^T w_m \end{bmatrix} ; \ c = -\Re \begin{bmatrix} \sum_{i=1,\dots,r} X_1(j\omega_i)G_{\text{ref}}(j\omega_i)^* e_1 \\ \vdots \\ \sum_{i=1,\dots,r} X_m(j\omega_i)G_{\text{ref}}(j\omega_i)^* e_m \end{bmatrix}$$

If $r > q$ the matrix H is positive definite. Note that $A_1 \in \mathbb{R}^{mpq \times m}$, $b_1 \in \mathbb{R}^{m \times 1}$, $H \in \mathbb{R}^{mpq \times mpq}$ and $c \in \mathbb{R}^{mpq \times 1}$

Proof. The proof of Lemma 4.1.4 is divided into two parts. First, the linear constraint (4.12) is transformed to the form "$\xi A_1 = b_1$", second the criterion of (4.16) is transformed to the form "$J = \xi H \xi^T - 2c\xi$"

Linear constraints. The eigenstructure assignment constraint (4.12) relative to the gain is $G(\lambda_i)(Cv_i + Dw_i) = w_i$. Let us consider this equation

$$G(\lambda_i)(Cv_i + Dw_i) = w_i \qquad (4.21)$$

Using (4.20), the kth row of (4.21) can be written as

$$e_k^T \sum_{j=1}^{m} e_j \xi_j X_j(\lambda_i)(Cv_i + Dw_i) = e_k^T w_i \ , \ k = 1, \ldots, m$$

Considering the fact that $e_k^T e_j = 1$ if $k = j$, $e_k^T e_j = 0$ otherwise, Equation (4.21) is equivalent to

$$\xi_k X_k(\lambda_i)(Cv_i + Dw_i) = e_k^T w_i \ , \ k = 1, \ldots, m \qquad (4.22)$$

which is of the form "$\xi A_1 = b_1$". Considering together the m rows of (4.21) leads to the result stated in the lemma.

The quadratic criterion. Ignoring the sum with respect to frequencies ω_i, this criterion is

$$J = \text{trace}(-G_{\text{ref}}(j\omega_i) + G(j\omega_i))(-G_{\text{ref}}(j\omega_i) + G(j\omega_i))^*$$

or (removing the constant term $\text{trace}(G_{\text{ref}}(j\omega_i)G_{\text{ref}}(j\omega_i)^*))$

$$J = -\text{trace}(G(j\omega_i)G_{\text{ref}}(j\omega_i)^*) -$$

$$\text{trace}(G_{\text{ref}}(j\omega_i)G(j\omega_i)^*) + \text{trace}(G(j\omega_i)G(j\omega_i)^*)$$

$$J = -2\Re \ \text{trace}(G(j\omega_i)G_{\text{ref}}(j\omega_i)^*) + \text{trace}(G(j\omega_i)G(j\omega_i)^*) \qquad (4.23)$$

Let us consider the second term (a simpler treatment can be applied to the first one), using (4.20):

$$\text{trace}(G(j\omega_i)G(j\omega_i)^*) = \text{trace}(\sum_{k=1}^{m} e_k \xi_k X_k(j\omega_i) \sum_{l=1}^{m} X_l(j\omega_i)^* \xi_l^T e_l^T)$$

after permutation

$$\text{trace}(G(j\omega_i)G(j\omega_i)^*) = \sum_{k=1\ldots m, l=1\ldots m} \text{trace}(\xi_k X_k(j\omega_i) X_l(j\omega_i)^* \xi_l^T e_l^T e_k)$$

Considering the fact that the trace of a scalar times $e_k e_l^T$ is zero if $k \neq l$:

$$\text{trace}(G(j\omega_i)G(j\omega_i)^*) = \sum_{k=1}^{m} \xi_k X_k(j\omega_i) X_k(j\omega_i)^* \xi_k^T$$

Finally, with the first term of (4.23), the criterion can be written:

$$J = \sum_{k=1}^{m} (\xi_k X_k(j\omega_i) X_k(j\omega_i)^* \xi_k^T - 2\Re \ \xi_k X_k(j\omega_i) G_{\text{ref}}(j\omega_i)^* e_k)$$

which is, taking into account the sum with respect to the frequencies:

$$J = \sum_{k=1}^{m} \xi_k (\sum_{i=1}^{r} X_k(j\omega_i) X_k(j\omega_i)^*) \xi_k^T -$$

$$2\Re \sum_{k=1}^{m} \xi_k (\sum_{i=1}^{r} X_k(j\omega_i) G_{\text{ref}}(j\omega_i)^* e_k)$$

$$(4.24)$$

that is of the form "$\xi H \xi^T + 2\xi c$" as required for LQ-programming. This criterion is positive. For its definiteness, taking into account the special form of the matrix $X_k(\lambda)$ and the properties of Van der Monde determinants, at least $q+1$ frequencies must be considered in the sum of (1.86) $(\rightarrow r > q)$. ∎

Equation (4.24) explains a numerical problem that appears in some cases: the matrix $X_k(j\omega_i) X_k(j\omega_i)^*$ can be very badly conditioned. For example if $\omega_i = 100$ and $q = 5$, this matrix contains entries ranging from 1 to 10^{20}. That is why it is recommended in the toolbox to consider low denominator degrees and to consider the criterion frequencies close to the unity.

4.2   First order perturbations

This section treats all first order approximations considered in this book. First, is treated the case where the feedback gain is a constant matrix, and from §4.2.3 the transfer matrix case will is considered.

4.2.1   Proof of Lemmas 1.2.2 and 1.4.1

This lemma is stated on page 34. It is recalled here with some additional details. The system (4.1) is assumed to be controlled by a proportional feedback K and $w_i = K(Cv_i + Dw_i)$ and $t_i = (u_iB + t_iD)K$.

LEMMA 4.2.1 *The closed-loop left and right eigenvectors corresponding to the eigenvalue λ_i are denoted u_i and v_i (assumed to be normalized such that $u_iv_i = 1$). For a variation of the closed-loop state-space matrix $\widehat{A} = A + B(I - KD)^{-1}KC$ we have at the first order:*

$$\Delta\lambda_i = u_i\Delta\widehat{A}\ v_i \qquad (4.25)$$

Moreover if the variation $\Delta\widehat{A}$ of the matrix $\widehat{A}$ is due to a variation ΔK of the gain K:

$$\Delta\lambda_i = (u_iB + t_iD)\Delta K(Cv_i + Dw_i) \qquad (4.26)$$

and if the variation $\Delta\widehat{A}$ is due to the variations ΔA, ΔB, ΔC, ΔD of the matrices A, B, C, D:

$$\Delta\lambda_i = u_i\Delta Av_i + u_i\Delta Bw_i + t_i\Delta Cv_i + t_i\Delta Dw_i \qquad (4.27)$$

Proof. In order to check Equation (4.25) (see [Wilkinson, 1965]), let us differentiate $\widehat{A}\ v_i = \lambda_iv_i$.

$$(\hat{A} + \Delta\hat{A})(v_i + \Delta v_i) = (\lambda_i + \Delta\lambda_i)(v_i + \Delta v_i)$$

multiplying on the left by u_i, note that $u_iv_i = 1$ and $u_i\hat{A} = \lambda_iu_i$:

$$u_i\Delta\hat{A}v_i + u_i\Delta\hat{A}\Delta v_i = \Delta\lambda_i + u_i\Delta\lambda_i\Delta v_i$$

The first order approximation of this equation is equivalent to:

$$u_i\Delta\hat{A}v_i = \Delta\lambda_i$$

So, Equation (4.25) is satisfied. In Equation (4.26), the variation of $\widehat{A}$ is due to a feedback gain variation (ΔK):

$$\Delta\hat{A} = B(I - KD)^{-1}\Delta KC + BK(I - KD)^{-1}\Delta KD(I - KD)^{-1}KC$$

So, Equation (4.25) is equivalent to:

$$\Delta\lambda_i = \left(u_i B(I - KD)^{-1}\right) \Delta K \left(Cv_i + D(I - KD)^{-1}KC\right) v_i$$

From the definition of w_i, $w_i = (I - KD)^{-1}KCv_i$, therefore, the right hand side parenthesis is equal to $Cv_i + Dw_i$. The left hand side parenthesis is treated considering the third matrix identity of page 247 that is $(I - KD)^{-1} = I + K(I - DK)^{-1}D$ and the definition of t_i: $t_i = u_i BK(I - DK)^{-1}$. It follows that the left hand side parenthesis can be written $(u_i B + t_i C)$, so, the result stated in Equation (4.26) is clearly satisfied. In Equation (4.27) it is the matrices A, B, C, D that vary. Let us write

$$\Delta\widehat{A} = \Delta A + \Delta BK(I - DK)^{-1}C +$$

$$BK(I - DK)^{-1}\Delta C + BK(I - DK)^{-1}\Delta DK(I - DK)^{-1}C$$

Using again (1.3) and (1.4):

$$u_i\Delta\widehat{A}v_i = u_i\Delta Av_i + u_i\Delta Bw_i + t_i\Delta Cv_i + t_i\Delta Dw_i$$

So, Equation (4.27) is satisfied. ∎

4.2.2   Proof of Lemma 1.4.2

For avoiding cumbersome notations, it is assumed that $D = 0$. See the comment given after the proof for the general case ($D \neq 0$). The following matrix notations will be used:

$$V_i = \begin{bmatrix} v_1 & \cdots & v_{i-1} & v_{i+1} & \cdots & v_n \end{bmatrix}$$

$$U_i^T = \begin{bmatrix} u_1^T & \cdots & u_{i-1}^T & u_{i+1}^T & \cdots & u_n^T \end{bmatrix}$$

$$\Lambda_i = \text{Diag}\{\lambda_1,\ldots,\lambda_{i-1},\lambda_{i+1},\ldots,\lambda_n\}$$

The vectors are normalized as follows $\|v_i\| = 1$ and $u_i v_i = 1$. Therefore:

$$U_i V_i = I_{n-1} \; ; \; V_i U_i = I_n - v_i u_i \; ; \; AV_i = V_i\Lambda_i \tag{4.28}$$

Consider the matrices U and V of the n left and right eigenvectors (column v_i added to V_i and row u_i added to U_i). From the above normalizations,

$$VU = I$$

but,

$$VU = V_i U_i + v_i u_i$$

so

$$V_i U_i = I - v_i u_i \qquad (4.29)$$

LEMMA 4.2.2 *The variation of eigenvectors, input and output directions due to the variations of the gain ΔK are as follows:*

Variations of right eigenvectors:

$$\begin{aligned}
\Delta v_i &= X_v \Delta K Y_v \quad \text{where} \\
X_v &= (I - v_i v_i^*) V_i (\lambda_i I - \Lambda_i)^{-1} U_i B \\
Y_v &= C v_i
\end{aligned} \qquad (4.30)$$

Variations of input directions:

$$\begin{aligned}
\Delta w_i &= X_w \Delta K Y_v \quad \text{where} \\
X_w &= I + KC X_v
\end{aligned} \qquad (4.31)$$

Variations of left eigenvectors:

$$\begin{aligned}
\Delta u_i &= X_u \Delta K Y_u - u_i X_v \Delta K Y_v u_i \quad \text{where} \\
X_u &= u_i B \\
Y_u &= C V_i (\lambda_i I - \Lambda_i)^{-1} U_i
\end{aligned} \qquad (4.32)$$

Variations of output directions:

$$\begin{aligned}
\Delta t_i &= X_u \Delta K Y_t - u_i X_v \Delta K Y_v t_i \quad \text{where} \\
Y_t &= I + Y_u B K
\end{aligned} \qquad (4.33)$$

Proof: *Variations of right eigenvectors.* Let us write

$$(A + B\Delta K C)(v_i + \Delta v_i) = (\lambda_i + \Delta \lambda_i)(v_i + \Delta v_i)$$

retaining the first order terms:

$$A\Delta v_i + B\Delta K C v_i = \lambda_i \Delta v_i + \Delta \lambda_i v_i$$

that is, using (4.25): $\Delta \lambda_i = u_i B \Delta K C v_i$

$$A\Delta v_i - \lambda_i \Delta v_i = v_i u_i B \Delta K C v_i - B\Delta K C v_i$$

consequently

$$(\lambda_i I - A)\Delta v_i = (I - v_i u_i) B \Delta K C v_i$$

Using (4.29) and (4.28), it can be checked that $\Delta v_i = V_i(\lambda_i I - \Lambda_i)^{-1} U_i B \Delta K C v_i$ satisfies the above equation. Noting that v_i generates the kernel of $(\lambda_i I - A)$, all the solutions are given by

$$\Delta v_i = V_i(\lambda_i I - \Lambda_i)^{-1} U_i B \Delta K C v_i + \alpha v_i$$

(that is Equation (10.2) of [Wilkinson, 1965], written in a matrix form, but in this reference, the coefficient α is equal to zero). The parameter α will be chosen in such a way that the norm of v_i remains constant: $\Delta v_i^* v_i + v_i^* \Delta v_i = 0$, or more simply

$$v_i^* \Delta v_i = 0$$

with $v_i^* v_i = 1$, the required value of α can be derived leading to:

$$\Delta v_i = (I - v_i v_i^*) V_i (\lambda_i I - \Lambda_i)^{-1} U_i B \Delta K C v_i$$

Summary:

$$\begin{aligned}
\Delta v_i &= X_v \Delta K Y_v \text{ where} \\
X_v &= (I - v_i v_i^*) V_i (\lambda_i I - \Lambda_i)^{-1} U_i B \\
Y_v &= C v_i
\end{aligned} \tag{4.34}$$

Variations of input directions. Input directions are given by Equation (1.3), that is, $w_i = K C v_i$, then

$$\Delta w_i = \Delta K C v_i + K C \Delta v_i$$

that is,

$$\begin{aligned}
\Delta w_i &= X_w \Delta K Y_v \text{ where} \\
X_w &= I + K C X_v
\end{aligned} \tag{4.35}$$

Variations of left eigenvectors. As for the right eigenvectors case, we obtain

$$\Delta u_i = u_i B \Delta K C V_i (\lambda_i I - \Lambda_i)^{-1} U_i + \beta u_i$$

where β is the parameter to be chosen for normalization. The normalization is here $u_i v_i = 1$, so,

$$\Delta u_i v_i + u_i \Delta v_i = 0$$

then

$$\beta = -u_i X_v \Delta K Y_v$$

Summary:

$$\begin{aligned}
\Delta u_i &= X_u \Delta K Y_u - u_i X_v \Delta K Y_v u_i \text{ where} \\
X_u &= u_i B \\
Y_u &= C V_i (\lambda_i I - \Lambda_i)^{-1} U_i
\end{aligned} \tag{4.36}$$

Variations of output directions. Output directions are given by (1.4), that is, $t_i = u_i B K$ then

$$\Delta t_i = \Delta u_i B K + u_i B \Delta K$$

so,

$$\begin{aligned}
\Delta t_i &= X_u \Delta K Y_t - u_i X_v \Delta K Y_v t_i \quad \text{where} \\
Y_t &= I + Y_u B K
\end{aligned} \qquad (4.37)$$

∎

Comment. In the general case $D \neq 0$, it suffices to replace K by $K(I - DK)^{-1}$ in (4.30) - (4.33). This is a consequence of the well known result that states that, applying the feedback gain $\widetilde{K} = K(I - DK)^{-1}$ to the system (A, B, C) is equivalent to applying the feedback gain K to the system (A, B, C, D) (see the comments of page 247).

4.2.3   Proof of Lemma 1.3.2

Lemma 1.3.2 gives the direction in the complex plane of the variations of the open-loop poles when the loop gain (δ in Figure 1.12, page 70) varies from zero.

LEMMA 4.2.3 *Consider System (4.1) controlled by a dynamic feedback $\delta G_0(s)$ where δ is a small real number. The open-loop left and right eigenvectors of A corresponding to the eigenvalue λ_i are denoted u_i and v_i (assumed to be normalized such that $u_i v_i = 1$).*

$$\Delta \lambda_i = \delta u_i B G_0(\lambda_i) C v_i \qquad (4.38)$$

is the first order approximation of the variation of λ_i induced by the variation of δ.

Proof. Let (A_c, B_c, C_c, D_c) denote the state-space representation of $G_0(s)$. The cascade connection of $G_0(s)$ and of System (4.1) leads to the following sate-space representation

$$\begin{bmatrix} \dot{x} \\ \dot{x}_c \end{bmatrix} = \begin{bmatrix} A & BC_c \\ 0 & A_c \end{bmatrix} \begin{bmatrix} x \\ x_c \end{bmatrix} + \begin{bmatrix} BD_c \\ B_c \end{bmatrix} u_c \qquad (4.39)$$

$$y = \begin{bmatrix} C & DC_c \end{bmatrix} \begin{bmatrix} x \\ x_c \end{bmatrix} + DD_c u_c \qquad (4.40)$$

y is connected to u_c by a non-dynamic feedback of the form δI. The result of Lemma 4.2.1 will be applied considering that the feedback variation is δI. But before, it is necessary to compute the left and right eigenvectors of System (4.39). Let λ_i be an eigenvalue of A and v_i, u_i denote the corresponding left and right eigenvectors. It is straightforward to check that

$$\begin{bmatrix} A & BC_c \\ 0 & A_c \end{bmatrix} \begin{bmatrix} v_i \\ 0 \end{bmatrix} = \lambda_i \begin{bmatrix} v_i \\ 0 \end{bmatrix}$$

and

$$[u_i \quad -u_i BC_c(A_c - \lambda_i I)^{-1}] \begin{bmatrix} A & BC_c \\ 0 & A_c \end{bmatrix} = \lambda_i [u_i \quad -u_i BC_c(A_c - \lambda_i I)^{-1}]$$

Applying the result of Lemma 1.2.2 to the considered problem leads to:

$$\Delta\lambda = [u_i \quad - u_i BC_c(A_c - \lambda_i I)^{-1}] \begin{bmatrix} BD_c \\ B_c \end{bmatrix} \delta I \, [\, C \quad DC_c \,] \begin{bmatrix} v_i \\ 0 \end{bmatrix}$$

that is,

$$\Delta\lambda = \delta \; u_i B(C_c(\lambda_i I - A_c)^{-1}B_c + D_c)Cv_i \; \blacksquare$$

4.2.4   Linear Quadratic Programming

This paragraph addresses the problem of writing design specifications related to Equation (4.38) as a Linear Quadratic Programming problem. Some relevant practical inequalities related to (4.38) are proposed on page 71. Only two of these inequalities that capture all possibilities are considered now:

- $\delta_{Ii}\Re(\Delta(\lambda_i)) \le \delta_{Ri}\Im(\Delta(\lambda_i))$

- $\Re(\Delta(\lambda_i)) \le \delta_{Ri}$

In view of Lemma 4.2.3 these inequalities can be written respectively:

- $\delta_{Ri}\Re(u_i BG_0(\lambda_i)Cv_i) \le \delta_{Ii}\Im(u_i BG_0(\lambda_i)Cv_i)$

- $\Re(u_i BG_0(\lambda_i)Cv_i) \le \delta_{Ri}$

The following lemma gives the constraints in a form that is compatible with Lemma 4.1.4. The transfer matrix form of $G_0(s)$ is as in Equation (4.15). The denominators are fixed and the notations (4.17) and (4.19) are used.

LEMMA 4.2.4 *The inequality relative to $G_0(s)$ of the form :*

$$\delta_{Ri}\Re(u_i BG_0(\lambda_i)Cv_i) \le \delta_{Ii}\Im(u_i BG_0(\lambda_i)Cv_i)$$

or respectively

$$\Re(u_i BG_0(\lambda_i)Cv_i) \le \delta_{Ri}$$

together with a quadratic criterion as in (4.24)) is a Linear Quadratic Programming problem $\xi A_2 \le b_2$, $J = \xi H\xi^T + 2\xi c$ defined by :

$$A_2 = \begin{bmatrix} \delta_{Ri}\Re(u_i Be_1 X_1(\lambda_i)Cv_i) - \delta_{Ii}\Re(u_i Be_1 X_1(\lambda_i)Cv_i) \\ \vdots \\ \delta_{Ri}\Re(u_i Be_m X_m(\lambda_i)Cv_i) - \delta_{Ii}\Re(u_i Be_m X_m(\lambda_i)Cv_i) \end{bmatrix} \; ; \; b_2 = 0$$

or respectively

$$A_2 = \begin{bmatrix} \Re(u_i Be_1 X_1(\lambda_i)Cv_i) \\ \vdots \\ \Re(u_i Be_m X_m(\lambda_i)Cv_i) \end{bmatrix} \; ; \; b_2 = \delta_{Ri}$$

H and c being as specified in Proposition 4.1.4.

Proof. Consider:

$$\delta_{Ri}\Re(u_iBG_0(\lambda_i)Cv_i) \le \delta_{Ii}\Im(u_iBG_0(\lambda_i)Cv_i)$$

$G_0(s)$ being written like $G(s)$ in (4.15) - (4.20),

$$\delta_{Ri}\Re(\sum_{k=1}^m u_iBe_k\xi_kX_k(\lambda_i)Cv_i) \le \delta_{Ii}\Im(\sum_{k=1}^m u_iBe_k\xi_kX_k(\lambda_i)Cv_i)$$

as u_iBe_k are scalars and ξ_k are real vectors,

$$\sum_{k=1}^m \xi_k\delta_{Ri}\Re(u_iBe_kX_k(\lambda_i)Cv_i) \le \sum_{k=1}^m \xi_k\delta_{Ii}\Im(u_iBe_kX_k(\lambda_i)Cv_i)$$

which is equivalent to the linear inequality constraint:

$$[\xi_1 \quad \cdots \quad \xi_m] \begin{bmatrix} \delta_{R1}\Re(u_iBe_1X_1(\lambda_i)Cv_i) - \delta_{I1}\Re(u_iBe_1X_1(\lambda_i)Cv_i) \\ \vdots \\ \delta_{Rm}\Re(u_iBe_mX_m(\lambda_i)Cv_i) - \delta_{Im}\Re(u_iBe_mX_m(\lambda_i)Cv_i) \end{bmatrix} \le 0$$

$$(4.41)$$

The formulation of the quadratic criterion is the same as in the proof of Proposition 4.1.4. The second inequality can be proved more easily. ∎

4.3 Minimum energy assignment

Solving the Linear Quadratic problem consists of finding the sate feedback that minimizes the following criterion

$$J = \int_0^\infty (x^T Q x + u^T R u) dt$$

in which Q and R are positive symmetric matrices of appropriate size, in addition, R is definite. The minimum energy problem corresponds to a special form in which only the input vector u is weighted in the criterion *i.e.*, $Q = 0$ (and for simplifying notation $R = I$).

$$J = \int_0^\infty u^T u \, dt$$

The eigenstructure assignment interpretation of this problem is given in Lemma 1.1.3 page 24. This lemma states:

LEMMA 4.3.1 *The minimum energy eigenstructure assignment corresponds to*

- *preserving the open-loop pairs of eigenvalue/eigenvector for stable open-loop eigenvalues*

- *assigning the symmetric stable images (symmetric with respect to the imaginary axis) of the unstable open-loop eigenvalues. In that case the assigned right eigenvectors are given by:*

$$v_i = (A - \lambda_i I)^{-1} B B^T u_{0i}^*$$

where λ_i is the symmetric stable image of the open-loop eigenvalue λ_{0i} and u_{0i} is the left eigenvector of A corresponding to λ_{0i}.

4.3.1 The Hamiltonian solution to the Linear Quadratic Problem

The solution of the LQ problem is usually computed by considering the Hamiltonian matrix H defined by

$$H = \begin{bmatrix} A & -BR^{-1}B^T \\ -Q & -A^T \end{bmatrix}$$

For computing the optimal sate feedback, the stable eigenvalues and eigenvectors of this matrix are computed. Under mild assumptions, it can be shown that

- the eigenvalues of the matrix H are symmetric with respect to the imaginary axis

- the n stable eigenvalues of H are the assigned closed-loop eigenvalues. Let

Let $V_1 \in \mathbb{C}^{n \times n}$ and $V_2 \in \mathbb{C}^{n \times n}$ be the upper and lower parts of the matrix the columns of which are the eigenvectors corresponding to the n stable eigenvalues.

- V_1 is invertible.

- the optimal state feedback is given by $K = -R^{-1}B^T V_2 V_1^{-1}$.

This result will be used in the form:

$$KV_1 = -R^{-1}B^T V_2 \tag{4.42}$$

Now, is considered the minimum energy case ($Q = 0$, $R = I$). In this case, the eigenvalues of H are those of A and $-A^T$.

4.3.2 Proof of Lemma 1.1.3

The eigenstructure assignment constraints relative to K induced by the minimum energy problem are now treated separately for each eigenvalue.

Stable eigenvalues: λ_{0i} is a stable eigenvalue of A with corresponding right eigenvector v_{0i}. As λ_{0i} is also a stable eigenvalue of H, there exists two vectors v_{1i} and v_{2i} such that

$$H = \begin{bmatrix} A & -BRB^T \\ 0 & -A^T \end{bmatrix} \begin{bmatrix} v_{1i} \\ v_{2i} \end{bmatrix} = \lambda_i \begin{bmatrix} v_{1i} \\ v_{2i} \end{bmatrix} \tag{4.43}$$

From this equation, it can be concluded that $v_{1i} = v_{0i}$ and $v_{2i} = 0$. It remains to prove that the feedback K of (4.42) assigns this right eigenvector. Clearly from (4.42)

$$Kv_{1i} = -B^T v_{2i}$$

that is

$$Kv_{0i} = 0$$

so,

$$(A + BK)v_{0i} = Av_{0i} = \lambda_{0i}v_{0i}$$

So, the minimum energy problem consists of preserving open-loop stable eigenvalues and eigenvectors.

Unstable eigenvalues: λ_{0i} is an unstable eigenvalue of A with corresponding left eigenvector u_{0i}. The assigned eigenvalue is the stable image of

λ_{0i}, it is denoted λ_i. The corresponding eigenvector of H satisfies (4.43) that can be written:

$$(A - \lambda_i I)v_{1i} - BB^T v_{2i} = 0$$
$$(-A^T - \lambda_i I)v_{2i} = 0$$

The second equation means that v_{2i} is the left eigenvector of the matrix $-A$ corresponding to the λ_{0i}, so

$$v_{2i} = u^*_{0i} \qquad (4.44)$$

The first equation leads to

$$v_{1i} = (A - \lambda_i I)^{-1} BB^T u^*_{0i}$$

this equation will be used in the following form

$$(A - \lambda_i I)v_{1i} = BB^T u^*_{0i} \qquad (4.45)$$

It remains to prove that the feedback K of (4.42) assigns v_{i1} as right eigenvector. Clearly from (4.42)

$$Kv_{1i} = -B^T v_{2i}$$

so,

$$(A + BK)v_{1i} = Av_{1i} - BB^T v_{2i}$$

that becomes after substitution of (4.44) and (4.45)

$$(A + BK)v_{1i} = Av_{1i} - (A - \lambda_i I)v_{1i}$$

$$(A + BK)v_{1i} = \lambda_i v_{1i}$$

which is as stated in the lemma. $\blacksquare$

4.4   Pole assignment by output feedback

Most of the existing pole assignment techniques can be viewed as special cases of the algorithm framework presented in [Magni and Champetier, 1991]. But there are not many algorithms in this framework for which all degrees of freedom ($m \times p$ that is equal to the number gain coefficients for a system with p outputs and m inputs) are really used for pole assignment. In most cases, a part of the degrees of freedom is *a priori* lost for obtaining simpler "sub-problems". Even in the state feedback case such a situation is encountered, for example all techniques using the Brunovski canonical form ([Wonham, 1979] belong to this class of techniques (except in the single input case). It is clear that all techniques that do not exhibit all degrees of freedom must be rejected for practical applications. For this reason, only one technique was retained in this toolbox. This technique is described in Lemma 1.2.1 page 33 (that is Lemma 4.4.1 below).

It is assumed that $D = 0$, however, if $D \neq 0$ it suffices to compute the gain assuming that $D = 0$ and then to replace the result by $K(I-DK)^{-1}$. Lemma 1.2.1 is recalled:

LEMMA 4.4.1 *Let us consider q_1 triples (λ_i, v_i, w_i), $q_1 < p$ satisfying (4.4). and an additional non-zero vector v_{q_1+1} satisfying*

$$v_{q_1+1} \in \text{Im } V(\lambda_{q_1+1}) \cap ((A - \beta_1 I)\text{Ker}C + \text{Im } [v_1 \ldots v_{q_1}]) \cap \ldots$$

$$\cap ((A - \beta_{q_2} I)\text{Ker}C + \text{Im } [v_1 \ldots v_{q_1}]) \qquad (4.46)$$

Then, all feedback gain K such that

$$KC[v_1 \ldots v_{q_1} \ v_{q_1+1}] = [w_1 \ldots w_{q_1} \ w_{q_1+1}] \qquad (4.47)$$

assigns the $q_1 + 1$ triples (λ_i, v_i, w_i), $i = 1, \ldots, q_1 + 1$ and assigns the additional q_2 eigenvalues $\beta_1, \ldots, \beta_{q_2}$.

A condition relative to q_1 and q_2 such that the intersection in Equation (4.46) does not reduce to zero will be discussed later.

4.4.1   Technical preliminary results

LEMMA 4.4.2 *Let $V = [v_1 \ldots v_q]$ and $W = [w_1 \ldots w_q]$ be such that (λ_i, v_i, w_i) satisfy (4.4). It is assumed that $q \leq p$ and*

$$\text{rank}(CV) = q$$

Then all feedback gain K satisfying

$$KCV = W$$

assigns $\lambda_i, \ldots, \lambda_q$ plus the zeros of the triple (A, V, C)

Proof. The fact that $\{\lambda_i, \ldots, \lambda_q\}$ are assigned, results from Lemma 4.1.1. From the rank condition relative to CV, β is a zero of the triple (A, V, C) if and only if there exists a non-zero vector x such that (see [MacFarlane and Karcanias, 1976])

$$\begin{bmatrix} A - \beta I & V \\ C & 0 \end{bmatrix} \begin{bmatrix} x \\ \xi \end{bmatrix} = 0 \tag{4.48}$$

So, we have

$$Ax = \beta x - V\xi \quad \text{and} \quad Cx = 0$$

therefore ($[x \ \ V]$ is a maximal rank matrix since $\text{Im}V \cap \text{Ker}C = 0$ and $x \in \text{Ker}C$)

$$(A + BKC)[\,x \ \ V\,] = [\,x \ \ V\,] \begin{bmatrix} \beta & 0 \\ -\xi & \Lambda \end{bmatrix}$$

which means that the zero β belongs to the closed-loop spectrum. ∎

LEMMA 4.4.3 *Assume that V is as in Lemma 4.4.2 and* $\text{rank}(CV) = q \leq p$, β *is a zero of (A, V, C) if and only if*

$$\text{Im}V \cap (A - \beta_i I)\text{Ker}C \neq 0$$

Proof. If β is a zero of (A, V, C), from (4.48), there exists non-zero vectors[2] x and ξ such that

$$\begin{bmatrix} A - \beta I \\ C \end{bmatrix} x = - \begin{bmatrix} V \\ 0 \end{bmatrix} \xi$$

which means that the vector $V\xi$ belongs to $(A - \beta_i I)\text{Ker}C$. The proof of the converse is similar. ∎

4.4.2   Proof of Lemma 1.2.1

Assuming that the intersection of Equation (4.46) does not reduce to zero, for a non zero vector v_{q_1+1} we have

$$v_{q_1+1} \in \text{Im } V(\lambda_{q_1+1}) \cap ((A - \beta_i I)\text{Ker}C + \text{Im } [v_1 \ldots v_{q_1}])$$

so,

$$(A - \beta_i I)\text{Ker}C \cap \text{Im } [v_1 \ldots v_{q_1+1}]) \neq 0$$

which means from Lemma 4.4.3 that assigning $[v_1 \ldots v_{q_1+1}]$ induces the assignment of β_i. This result is also true for $i = 1, \ldots, q_2$. The proof of Lemma 1.2.1 follows immediately. ∎

It remains to discuss the choice of q_1 and q_2 so that the intersection of Equation (4.46) does not reduce to zero. There are numerous special cases that are not easy to exhibit because the dimension of the intersection is closely related to the system observability indices (see [Champetier and Magni, 1991]), so it is suggested to use trials and errors. Regardless of these indices, it can easily be checked that the generic dimension of the intersection is equal to $m + q_2(q_1 - p)$. q_1 and q_2 must be chosen such that this value is strictly positive.

A typical application of this pole assignment technique concerns systems satisfying $m + p > n$, in which case, it is suggested to choose

- $q_1 = p - 1$

- $q_2 = n - p$

In this case, the generic dimension of the intersection is $m + (n - p)(p - 1 - p) = m + p - n$ that is strictly positive. It remains to count the degrees of freedom that are really used:

- q_1 eigenvalues and eigenvectors in m-dimensional subspaces, so mq_1 degrees of freedom

- one eigenvalue and one eigenvector in the above intersection, so $m + p - n$ degrees of freedom

- q_2 eigenvalues.

The total number of degrees of freedom is $mq_1 + m + p - n + q_2 = mp$ that is exactly equal to the number of entries of the matrix K, therefore the technique of Lemma 1.2.1 exhibits all degrees of freedom.

4.5   Non-interactive control design

This appendix justifies the design procedure that is stated in §1.2.4. For that purpose, the traditional technique of [Falb and Wolovich, 1967] is recalled without proof. It will be shown that this traditional approach and the proposed eigenstructure assignment technique are equivalent.

A square system, *i.e.*, a system such that the dimension of the vector of regulated outputs z is equal to the dimension of the input vector u are equal (to m)

$$\dot{x} = Ax + Bu$$
$$z = Ex$$

This system is assumed to be controllable.

Problem definition. We look for a state feedback control law

$$u = Kx + Hz_R$$

such that there is no cross-coupling between the entries of the reference vector z_R and of the regulated output vector z. In addition, the steady state gain between z_{Ri} and z_i must equal to 1. The traditional approach for decoupling is first recalled.

4.5.0.1   The traditional approach for decoupling

Definition of n_i. E_i denotes the ith row of the matrix E. The integers n_i $i = 1, \ldots, m$ are defined as follows:

$$E_iB = 0, \ E_iAB = 0, \ldots, \ E_iA^{n_i-1}B = 0, \ E_iA^{n_i}B \neq 0$$

Design Procedure 1. See [Falb and Wolovich, 1967] for a justification.
1 - Compute the integers n_i, $i = 1, \ldots, m$
2 - Select the polynomials $P_i(s)$, $i = 1, \ldots, m$ corresponding to the desired dynamics between z_{Ri} and z_i *i.e.*,

$$z_i = \frac{P_i(0)}{P_i(s)} z_{Ri}$$

where the degree of P_i is equal to $n_i + 1$, and where the leading coefficient of P_i is the unity (monic polynomial).
3 - If the matrix

$$\begin{bmatrix} E_1 A^{n_1} B \\ \vdots \\ E_m A^{n_m} B \end{bmatrix} \tag{4.49}$$

is non-singular, compute the feedback gain

$$K = - \begin{bmatrix} E_1 A^{n_1} B \\ \vdots \\ E_m A^{n_m} B \end{bmatrix}^{-1} \begin{bmatrix} E_1 P_1(A) \\ \vdots \\ E_m P_m(A) \end{bmatrix} \tag{4.50}$$

4 - Compute the feedforward gain

$$H = \begin{bmatrix} E_1 A^{n_1} B \\ \vdots \\ E_m A^{n_m} B \end{bmatrix}^{-1} \begin{bmatrix} P_1(0) & & 0 \\ & \ddots & \\ 0 & & P_m(0) \end{bmatrix} \tag{4.51}$$

Comments. Note that the sum $\sum_{i=1,\ldots,m}(n_i+1)$ is the number of assigned poles. If this sum is less than n, a question arises about the n_z non assigned poles.

$$n_z = n - \sum_{i=1,\ldots,m} (n_i + 1)$$

It can be shown that

- n_z is equal to the number of invariant zeros of the triple (A, B, E).

- The feedback K assigns the n_z invariant zeros as closed-loop poles.

Note that decoupling leads to an unstable solution if the zeros of (A, B, E) are in the left half complex plane.

4.5.0.2   Eigenstructure assignment approach

Let us first describe the design procedure. It will be shown that the matrices K and H that are defined in both design procedures are equal. Note that for simplifying notations, it is assumed that the invariant zeros of the triple (A, B, E) are all distinct.

Design Procedure 2.

1 - Compute the numbers n_i, $i = 1, \ldots, m$
2 - Choose m sets of $n_i + 1$ (self conjugate) eigenvalues that will be associated with the regulated output z_i. These sets are denoted

$$\{\lambda_{i1}, \ldots, \lambda_{i(n_i+1)}\}, i = 1, \ldots, m$$

Then, compute the vectors v_{ik} and w_{ik}, $i = 1, \ldots, m$, $k = 1, \ldots, n_i + 1$, such that

$$\begin{bmatrix} A - \lambda_{ik}I & B \\ E_1 & 0 \\ \vdots & \\ E_{i-1} & 0 \\ E_{i+1} & 0 \\ \vdots & \\ E_m & 0 \end{bmatrix} \begin{bmatrix} v_{ik} \\ w_{ik} \end{bmatrix} = 0$$

3 - Compute the n_z invariant zeros of (A, B, E) denoted z_i and the vectors v_{zi} and w_{zi} satisfying

$$\begin{bmatrix} A - z_iI & B \\ E & 0 \end{bmatrix} \begin{bmatrix} v_{zi} \\ w_{zi} \end{bmatrix} = 0$$

4 - Consider the matrices V the columns of which are all the vectors v_{ik} and v_{z_i}. The matrix W is built in a similar way (same ordering) form the vectors w_{ik} and w_{z_i}. Then compute K as follows

$$K = WV^{-1} \tag{4.52}$$

5 - Compute the matrix H

$$H = -(E(A + BK)^{-1}B)^{-1} \tag{4.53}$$

Before discussing further the advantages or limitations of this design procedure, some technical points are considered.

4.5.0.3   Technical results

Using the notations defined in both design procedures:

$$P_i(s) = \prod_{k=1}^{k=n_i+1} (s - \lambda_{ik})$$

LEMMA 4.5.1 *If $E_iB = 0, E_iAB = 0, \ldots, E_iA^{n_i-1}B = 0$ (and $E_iA^{n_i}B \neq 0$), then for all $j \neq i$*

$$E_iv_{jk} = 0, E_iAv_{jk} = 0, \ldots, E_iA^{n_i}v_{jk} = 0, E_iA^{n_i+1}v_{jk} = -E_iA^{n_i}Bw_{jk} \tag{4.54}$$

and concerning the zeros, $j = 1, \ldots, n_z$

$$E_iv_{zj} = 0, E_iAv_{zj} = 0, \ldots, E_iA^{n_i}v_{zj} = 0, E_iA^{n_i+1}v_{zj} = -E_iA^{n_i}Bw_{zj} \tag{4.55}$$

Proof. The proof is similar in both cases, so only (4.54) is proved. As $j \neq i$, $E_i v_{jk} = 0$ (from the design procedure). Also, from the design procedure

$$(A - \lambda_{jk}I)v_{jk} + Bw_{jk} = 0$$

Multiplying on the left by E_i

$$E_i A v_{jk} - \lambda_{jk} E_i v_{jk} + E_i B w_{jk} = 0$$

So,

$$E_i A v_{jk} = 0$$

By induction, multiplying of the left by $E_i A$, $E_i A^2$, and so on, it can be shown that $E_i A^2 v_{jk} = 0, \ldots, E_i A^{n_i} v_{jk} = 0$. At the last step, we multiply by $E_i A^{n_i}$:

$$E_i A^{n_i+1} v_{jk} - \lambda_{jk} A^{n_i} E_i v_{jk} + E_i A^{n_i} B w_{jk} = 0$$

So,

$$E_i A^{n_i+1} v_{jk} = -E_i A^{n_i} B w_{jk} \; \blacksquare$$

LEMMA 4.5.2 *If* $E_i B = 0, E_i AB = 0, \ldots, E_i A^{n_i-1} B = 0$ *(and* $E_i A^{n_i} B \neq 0$*), then*

$$E_i P_i(A) v_{ik} = -E_i A^{n_i} B w_{ik} \qquad (4.56)$$

Proof. From the definition of P_i,

$$E_i P_i(A) v_{ik} = E_i \left(\prod_{\substack{l=1 \\ l \neq k}}^{l=n_i+1} (A - \lambda_{il}I) \right) (A - \lambda_{ik}I)v_{ik}$$

From the design procedure $(A - \lambda_{ik}I)v_{ik} = -Bw_{ik}$, so,

$$E_i P_i(A) v_{ik} = -E_i \left(\prod_{\substack{l=1 \\ l \neq k}}^{l=n_i+1} (A - \lambda_{il}I) \right) Bw_{ik}$$

Expanding the above product and considering the fact that $E_i B = \ldots = E_i A^{n_i-1} B = 0$ we obtain

$$E_i P_i(A) v_{ik} = -E_i A^{n_i} B w_{ik} \; \blacksquare$$

LEMMA 4.5.3 *If the feedback gains of Equations (4.50) and (4.52) exist, they are equal. The feedforward gains of Equations (4.51) and (4.53) are equal.*

Proof. The feedback gains equality is equivalent to

$$
\begin{bmatrix} E_1 A^{n_1} B \\ \vdots \\ E_m A^{n_m} B \end{bmatrix}^{-1}
\begin{bmatrix} E_1 P_1(A) \\ \vdots \\ E_m P_m(A) \end{bmatrix} = -WV^{-1}
$$

that is equivalent to

$$
\begin{bmatrix} E_1 P_1(A) \\ \vdots \\ E_m P_m(A) \end{bmatrix} V = -
\begin{bmatrix} E_1 A^{n_1} B \\ \vdots \\ E_m A^{n_m} B \end{bmatrix} W
$$

From (4.54), (4.55) and (4.56) this equality is satisfied.

The feedforward gain of the second procedure is defined so that the steady state gain between z_R and z is the identity matrix. This is a necessary condition (decoupling in steady state), so, both feedforward gains are equal. ∎

Comments. In the second design procedure we ignored the problem of the invertibility of the matrix V. It can be shown that for almost any choice of the vectors v_{jk} the resulting matrix V is non-singular (provided that the number of poles associated with the ith regulated output is exactly equal to $n_i + 1$). However, the derivation of such a result is quite technical and is not really useful.

4.5.0.4 Discussion relative to the eigenstructure assignment approach

Discussion relative to the zeros. The modal interpretation of the problem of non-interactive control is summarized in Figure 1.3. Two kinds of decoupling properties must be considered

- the transmission of modes towards the regulated outputs

- the transmission of reference inputs towards the modes.

At first sight, Design Procedure 2 seems to ignore the second class of decoupling properties. But, under mild assumptions (invertibility of the matrix V and distinct zeros), Lemma 4.5.3 states the equivalence of the results obtained applying Design Procedures 1 and 2. Therefore, it is indirectly proved that the transmissions from reference signals towards the modes are as expected (see Figure 1.3).

Decoupling is the most popular application of eigenstructure assignment. Non-interactive flight control systems designed in that way are implemented in some civil and military aircrafts in current operation. The success of these applications seems to be in contradiction with the lack of available theoretical justifications. Indeed, the possible existence of zeros was mostly ignored applying Design Procedure 2. The illustrative examples of §1.2.4 show that ignoring the zeros seems at first sight to lead to acceptable results (feedback and feedforward gains having acceptable size) but simulations reveal very large transient cross coupling.

Approximate decoupling. Design Procedure 2 offers some flexibility for approximate decoupling:

- Output feedback case with as many measurements as dominant poles: In this case, the feedback is of the form $K = W(CV + DW)^{-1}$. If this gain does not perturb too much non-dominant poles, these poles remain non-dominant, therefore, can be ignored. However, the number of poles associated to each regulated output is equal to $n_i + 1$ minus some value related to the number of ignored poles[3].

- Output feedback with not enough measurements: In this case, an observer can be added to the system, the observer dynamics being uncontrollable (see Lemma 4.7.6) observer modes can be ignored.

- If the number of inputs is larger than the number of regulated outputs, the expression of the feedforward gain H of Equation (4.53) can be replaced by the solution with minimum norm of

$$-(E(A + BK)^{-1}B)H = I$$

- The "true" zeros analysis can be replaced by almost zeros analysis (see 1.5.3, page 102).

The resulting amended design procedure is the one adopted in §1.2.4.

4.6   Dynamic feedforward design

Let us consider the dynamic feedforward gain modeled in state-space form as

$$\begin{aligned}
\dot{x}_f &= A_f x_f + B_f v \\
y_f &= C_f x_f + v
\end{aligned} \tag{4.57}$$

This system is connected at the input of System (4.1) such that $u = y_f$. The resulting global system is:

$$\begin{cases}
\begin{bmatrix} \dot{x} \\ \dot{x}_f \end{bmatrix} = \begin{bmatrix} A & BC_f \\ 0 & A_f \end{bmatrix} \begin{bmatrix} x \\ x_f \end{bmatrix} + \begin{bmatrix} B \\ B_f \end{bmatrix} v \\[2ex]
\quad y \;\; = \begin{bmatrix} C & DC_f \end{bmatrix} \begin{bmatrix} x \\ x_f \end{bmatrix} + \quad D \quad v
\end{cases} \tag{4.58}$$

Proving Lemma 1.2.3 will consists of showing that some eigenvalues are cancelled and replaced by other ones in such a way that the corresponding eigenvectors can be selected as in the feedback case. In other words, the three following properties must be checked (pseudo-assignment by dynamic feedforward).

1 q *stable* eigenvalues of A denoted $\{\lambda_1 \ldots \lambda_q\}$ are made uncontrollable (*i.e.*, cancelled).

2 q new arbitrarily chosen eigenvalues $\{\beta_1 \ldots \beta_q\}$ are introduced.

3 For each introduced eigenvalue β_i, for any arbitrary vector v_i satisfying (4.4):

$$\begin{bmatrix} A - \beta_i I & B \end{bmatrix} \begin{bmatrix} v_i \\ w_i \end{bmatrix} = 0$$

there exists some vector v_{ci} such that

$$\begin{bmatrix} v_i \\ v_{ci} \end{bmatrix}$$

is an eigenvector of the global system (4.58).

In appearance, q pairs of eigenvalues / eigenvectors (β_i, v_i) are assigned replacing by cancellation q open-loop eigenvalues. The lower part of eigenvectors v_{ci} will be ignored for the same reason as on page 64.

First, a technical preliminary result is derived.

4.6.1   Rendering eigenvalues non controllable

This section addresses items 1 and 2 listed above. It is assumed that a set of q self-conjugate stable eigenvalues of the matrix A denoted

$\{\lambda_1 \ldots \lambda_q\}$ is selected. The corresponding left eigenvectors are denoted u_i.

$$U_q = \begin{bmatrix} u_1 \\ \vdots \\ u_q \end{bmatrix} \; ; \; \Lambda_q = \text{Diag}\{\lambda_1 \ldots \lambda_q\}$$

LEMMA 4.6.1 *The feedforward gain (A_f, B_f, C_f, I) satisfying*

$$B_f = -U_q B \quad and \quad A_f = \Lambda_q + B_f C_f \qquad (4.59)$$

is such that $\{\lambda_1, \ldots, \lambda_q\}$ become uncontrollable eigenvalues.

Proof: To render λ_i uncontrollable, we must choose (A_f, B_f, C_f) in such a way that a nonzero vector $[u_i \; u_{ci}]$ exists such that (see page 247)

$$\begin{bmatrix} u_i & u_{ci} \end{bmatrix} \begin{bmatrix} A - \lambda_i I & BC_f & B \\ 0 & A_f - \lambda_i I & B_f \end{bmatrix} = 0$$

So, we have three equations

$$u_i A = \lambda_i u_i \qquad (4.60)$$

$$u_i B C_f + u_{ci} A_f = \lambda_i u_{ci} \qquad (4.61)$$

$$u_i B + u_{ci} B_f = 0 \qquad (4.62)$$

From Equation (4.60), it is clear that the vectors u_i must be the left eigenvectors of A corresponding to the eigenvalues to be cancelled. With matrix notations (the definition of U_f is similar to the one of U_q), Equation (4.61) and (4.62) yield

$$U_q B C_f + U_f A_f = \Lambda_q U_f$$

$$U_q B + U_f B_f = 0$$

that can be written if U_f is invertible:

$$A_f = U_f^{-1} \Lambda_q U_f - B_f C_f$$

$$B_f = -U_f^{-1} U_q B$$

The matrices B_f, C_f, A_f and U_f are to be guessed from these equations. As there are too many degrees of freedom, it will be assumed that $U_f = I$, so the lemma is proved. ∎

Before continuing, it is interesting to give an interpretation of Lemma 4.6.1. This lemma states that for the required pole cancellation, B_f is fixed and A_f and C_f can be fixed simultaneously by designing C_f as a state feedback gain relative to the pair (Λ_f, B_f). In view of Equation (4.58), the poles assigned in that way are those we look for (denoted earlier $\{\beta_1, \ldots, \beta_q\}$). The next part of the derivation of the feedforward gain will consist of designing C_f as a feedback that assigns these eigenvalues but also, such that the corresponding eigenvectors are as required (*i.e.*, can be arbitrarily chosen satisfying (4.4)).

4.6.2   Eigenvector pseudo-assignment

The eigenvector pseudo-assignment by dynamic feedforward problem as stated in Lemma 1.2.3, is addressed in the next lemma.

LEMMA 4.6.2 *Assume that q triples (β_i, v_i, w_i) are selected, with $\beta_i \notin \sigma(A)$, and $\{U_q v_1, \ldots, U_q v_q\}$ linearly independent, satisfying*

$$\begin{bmatrix} A - \beta_i I & B \end{bmatrix} \begin{bmatrix} v_i \\ w_i \end{bmatrix} = 0 \qquad (4.63)$$

There exists a pair of matrices (A_f, C_f) satisfying (4.59) and

$$\begin{bmatrix} A - \beta_i I & BC_f \\ 0 & A_f - \beta_i I \end{bmatrix} \begin{bmatrix} v_i \\ v_{ci} \end{bmatrix} = 0 \qquad (4.64)$$

These matrices are given by

$$C_f = [w_1 \ldots w_q][v_{c1} \ldots v_{cq}]^{-1} \quad ; \quad A_f = \Lambda_q + B_f C_f \qquad (4.65)$$

where v_{ci} and w_i are defined by:

$$\begin{bmatrix} A - \beta_i I & 0 & B \\ 0 & \Lambda_q - \beta_i I & B_f \end{bmatrix} \begin{bmatrix} v_i \\ v_{ci} \\ w_i \end{bmatrix} = 0 \qquad (4.66)$$

Proof. The proposed proof of the lemma consists of checking that a pair of matrices (A_f, C_f) satisfying (4.65) and (4.66) satisfies also (4.64). Considering the bottom part of (4.66) that is

$$\begin{bmatrix} \Lambda_q - \beta_i I & B_f \end{bmatrix} \begin{bmatrix} v_{ci} \\ w_i \end{bmatrix} = 0$$

together with (4.65), it is clear that C_f is the state feedback gain relative to the pair (Λ_q, B_f) that assigns the triples (β_i, v_{ci}, w_i) (see Lemma 4.1.1). Therefore the bottom part of (4.64) is satisfied. Noting that from (4.65), $w_i = C_f v_{ci}$ the upper part of (4.64) is also satisfied.

It remains to show that the matrix $[v_{c1} \ldots v_{cq}]$ is invertible. From the definition of U_q (left eigenvectors):

$$\Lambda_q U_q = U_q A \tag{4.67}$$

Equation (4.63) is multiplied on the left by U_q

$$U_q A v_i - U_q \beta_i v_i + U_q B w_i = 0$$

After substitution of (4.59) and (4.67) this equation becomes

$$\Lambda_q U_q v_i - U_q \beta_i v_i - B_f w_i = 0$$

that can be written

$$(\Lambda_q - \beta_i I)(-U_q v_i) + B_f w_i = 0$$

which means that

$$v_{ci} = -U_q v_i \tag{4.68}$$

It follows from the hypothesis stated in the lemma that the matrix $[v_{c1} \ldots v_{cq}]$ is invertible. $\blacksquare$

4.6.3   Proof of Lemma 1.2.3

Lemma 1.2.3 is an alternative formulation of Lemma 4.6.2 organized as a design procedure. An additional condition relative to decoupling is considered. This condition is

$$\begin{bmatrix} E & 0 & F \end{bmatrix} \begin{bmatrix} v_i \\ v_{ci} \\ w_i \end{bmatrix}$$

that reduces to

$$\begin{bmatrix} E & F \end{bmatrix} \begin{bmatrix} v_i \\ w_i \end{bmatrix}$$

so, this constraint is an additional decoupling condition that just limits the possibilities for choosing vectors satisfying (4.63). This additional constraint is natural because in practice, it is rather natural to use dynamic feedforward for decoupling.

4.7   Observers

In this manual, observers are defined as banks of elementary observers. Elementary observer design can be viewed to some extent as the dual problem of a single right eigenvector assignment (compare (4.4) and (4.70)). Banks of elementary observers are similar to Luenberger observers, but such observers are usually designed in order to observe all the state components although in this manual, state observation is a secondary problem: Observers are often used in this book just in order to provide additional degrees of freedom for feedback design, the signals that are actually observed are ignored.

4.7.1   Elementary observers: Proof of Lemma 1.2.5

Lemma 1.2.5 is recalled here:

LEMMA 4.7.1 *The system defined by (see Figure 1.6):*

$$\dot{\hat{z}}_i = \pi_i \hat{z}_i - t_i y + u_i B u + t_i D u \tag{4.69}$$

where $u_i \in \mathbb{C}^n$, $t_i \in \mathbb{C}^p$ and $\pi_i \in \mathbb{C}$ satisfy :

$$\begin{bmatrix} u_i & t_i \end{bmatrix} \begin{bmatrix} A - \pi_i I \\ C \end{bmatrix} = 0 \tag{4.70}$$

is an observer of the variable $z_i = u_i x$ and the observation error $\epsilon_i = \hat{z}_i - u_i x$ satisfies:

$$\dot{\epsilon}_i = \pi_i \epsilon_i$$

Proof. Clearly from (4.1) and (4.69) we have:

$$\dot{\hat{z}}_i - u_i \dot{x} = \pi \hat{z}_i - t_i C x + u_i B u - u_i A x - u_i B u$$

hence, from (4.70) written as $u_i A + t_i C = \pi_i u_i$

$$\dot{\hat{z}}_i - u_i \dot{x} = \pi(\hat{z}_i - u_i x) \quad \blacksquare$$

Figure 1.6 (page 47) can be modified in such a way that the transfer matrix $S(s) = C(sI - A)^{-1}B + D$ appears in an explicit way, see Figure 4.1. The corresponding result is stated in the following lemma.

LEMMA 4.7.2 *The system defined by (see Figure 4.1)*

$$\dot{\hat{z}}_i = \pi_i \hat{z}_i + t_i (S(\pi_i) u - y) \tag{4.71}$$

where $t_i \in \mathbb{C}^p$ and $\pi_i \in \mathbb{C}$ with $\pi_i \notin \sigma(A)$ is an observer of the signal $z_i = u_i x$ where

$$u_i = t_i C(\pi_i I - A)^{-1} \tag{4.72}$$

and the observation error $\epsilon_i = \hat{z}_i - u_i x$ satisfies:

$$\dot{\epsilon}_i = \pi_i \epsilon_i$$

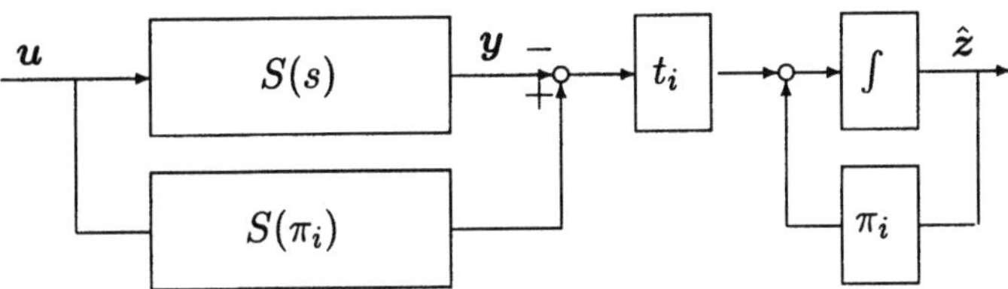

Figure 4.1. An elementary observer of $z_i = u_i x$ where u_i is given by (4.72).

Proof. Considering $S(\pi_i) = C(\pi_i I - A)^{-1} B + D$, we have

$$t_i(S(\pi_i)u - y) = t_i C(\pi_i I - A)^{-1} Bu + t_i Du - t_i y$$

Using (4.72)

$$t_i(S(\pi_i)u - y) = u_i Bu + t_i Du - t_i y$$

which is as in Lemma 1.2.5. ∎

The elementary observers presented in Lemmas 4.7.1 and 4.7.2 are equivalent (except condition $\pi_i \notin \sigma(A)$ of the second lemma) but are used differently. The first form is used for defining more general observers. The second form is interesting for implementation because if $S(s)$ is known as a function of some measured varying parameters, the observer can automatically be scheduled with respect to these parameters by replacing $S(\pi_i)$ by the parameter dependent form of $S(s)$ at $s = \pi_i$. This is particularly useful in the Linear Fractional Transformation representation framework (see [Magni et al., 1998] for more details).

Unknown input observers are treated on page 52. The main result is recalled in the following lemma.

LEMMA 4.7.3 *The observation error dynamic of the elementary observer (4.69) of System (4.1) subject to the unknown inputs d:*

$$\begin{aligned} \dot{x} &= Ax + Bu + E'd \\ y &= Cx + Du + F'd \end{aligned} \qquad (4.73)$$

is independent of d if u_i and t_i are selected such that

$$[u_i \quad t_i] \begin{bmatrix} A - \pi_i I & E' \\ C & F' \end{bmatrix} = 0 \qquad (4.74)$$

Proof. Clearly from (4.73) and (4.69) we have:

$$\dot{\hat{z}}_i - u_i\dot{x} = \pi\hat{z}_i - t_iCx + u_iBu - u_iAx - u_iBu - t_iF'd - u_iE'd$$

hence, from (4.74) written as $u_iA + t_iC = \pi_iu_i$

$$\dot{\hat{z}}_i - u_i\dot{x} = \pi(\hat{z}_i - u_ix) - (u_iE'_i + t_iF'_i)d$$

From (4.74), $u_iE'_i + t_iF'_i = 0$, therefore the lemma is proved. ∎

4.7.2 Observer with Kalman filter structure: Proof of Lemma 1.2.7

In this section and in the next ones, it will be assumed that a bank of q elementary observers is available. The corresponding vectors u_i and t_i and the dynamics π_i satisfy (4.70) which can be written in matrix form:

$$UA + TC = \Pi U \tag{4.75}$$

where

$$\Pi = \mathrm{Diag}\{\pi_1 \ldots \pi_q\} \; ; \; U = \begin{bmatrix} u_1 \\ \vdots \\ u_q \end{bmatrix} \; ; \; T = \begin{bmatrix} t_1 \\ \vdots \\ t_q \end{bmatrix} \tag{4.76}$$

In this section $q = n$. The observer with Kalman filter structure as in Figure 1.8 (page 51) is defined by

$$\begin{aligned} \dot{\hat{x}} &= A\hat{x} + Bu + K_o(\hat{y} - y) \\ \hat{y} &= C\hat{x} + Du \end{aligned} \tag{4.77}$$

in which K_o is the observer gain.

LEMMA 4.7.4 *The* state observer with Kalman filter structure *of Figure 1.8 (page 51) in which the gain K_o is computed as follows:*

- *n triples (π_i, u_i, t_i) satisfying (4.70) are selected*

- *$K_o = U^{-1}T$ where the matrices U and T are as defined in (1.59) (assuming that U is invertible),*

is equivalent to the observer of Figure 1.7 (page 49) in which

$$Q_y = 0 \quad and \quad Q_z = U^{-1} \tag{4.78}$$

Proof. From $K_o = U^{-1}T$ and $UA + TC = \Pi U$ (see (4.75)),

$$UK_o = T \quad \text{and} \quad U(A + K_oC) = \Pi U$$

Multiplying the first equation of (4.77) by U, replacing $\hat{y}$ by $C\hat{x} + Du$ and y by $Cx + Du$ we obtain

$$U\dot{\hat{x}} = U(A + K_oC)\hat{x} + UBu - UK_oCx$$

that is (using $U(A + K_oC) = \Pi U$)

$$U\dot{\hat{x}} = \Pi U\hat{x} + UBu - UK_o(y - Du)$$

that is (using $UK_o = T$)

$$\dot{\hat{z}} = \Pi\hat{z} + UBu - Ty + TDu$$

where $\hat{z} = U\hat{x}$. We recognize here the announced structure of Figure 1.7 in which Q_y and Q_z are as stated in (4.78). $\blacksquare$

4.7.3 Observer transfer function matrix: Proof of Lemma 1.2.6

The proof of Lemma 1.2.6 is divided into two parts (Lemmas 4.7.5 and 4.7.6).

LEMMA 4.7.5 *Considering the observer connected to the system as in Figure 1.7 (page 49), the transfer function matrix between u and $\hat{z}$ (see System (1.61)) is equal to $U(sI - A)^{-1}B$, which is also the transfer function matrix between u and z.*

Proof. Subtracting sU from both sides of (4.75)

$$U(A - sI) + TC = (\Pi - sI)U$$

Then multiply on the left by $(sI - \Pi)^{-1}$ and on the right by $(sI - A)^{-1}B$

$$(sI - \Pi)^{-1}UB - (sI - \Pi)^{-1}TC(sI - A)^{-1}B = U(sI - A)^{-1}B$$

which can be written using (4.2):

$$(sI - \Pi)^{-1}((UB + TD) - TS(s)) = U(sI - A)^{-1}B$$

the left transfer function matrix is the transfer from u to $\hat{z}$ and the right one is the transfer from u to z. $\blacksquare$

LEMMA 4.7.6 *Considering the observer connected to the system as in Figure 1.7, the eigenvalues of Π are uncontrollable poles.*

Proof. From Figure 1.7, the global state-space equation is

$$\begin{aligned}
\dot{x} &= Ax + Bu \\
\dot{\hat{z}} &= \Pi\hat{z} + (UB + TD)u - Ty \\
y &= Cx + Du
\end{aligned} \tag{4.79}$$

which can be written

$$\begin{bmatrix} \dot{x} \\ \dot{\hat{z}} \end{bmatrix} = \begin{bmatrix} A & 0 \\ -TC & \Pi \end{bmatrix} \begin{bmatrix} x \\ \hat{z} \end{bmatrix} + \begin{bmatrix} B \\ UB \end{bmatrix} u$$

In order to apply the controllability criterion of page 247, the corresponding "(A, B)" pair is considered

$$\begin{bmatrix} A - \pi_i I & 0 & B \\ -TC & \Pi - \pi_i I & UB \end{bmatrix}$$

This matrix is multiplied on the left by $[-u_i \ \ e_i]$ where e_i is a row vectors of zeros except the ith entry that is the unity

$$\begin{bmatrix} -u_i & e_i \end{bmatrix} \begin{bmatrix} A - \pi_i I & 0 & B \\ -TC & \Pi - \pi_i I & UB \end{bmatrix} =$$

$$\begin{bmatrix} -u_i(A - \pi_i I) - t_i C & e_i(\Pi_i - \pi_i I) & -u_i B + u_i B \end{bmatrix} = 0$$

So, π_i is not controllable. ∎

4.7.4   Separation Principle: Proof of Lemma 1.2.8

The Separation Principle is presented on page 57. It consists of using the following system

$$\begin{aligned}
\dot{x} &= Ax + Bu \\
\begin{bmatrix} y \\ z \end{bmatrix} &= \begin{bmatrix} C \\ U \end{bmatrix} x + \begin{bmatrix} D \\ 0 \end{bmatrix} u
\end{aligned} \tag{4.80}$$

instead of System (4.79) for feedback design. But the resulting feedback is applied to System (4.79).

LEMMA 4.7.7 Separation Principle *Assume that*

1 A n_cth order observer is designed, i.e. three matrices $U \in \mathbb{R}^{n_c \times n}$, $T \in \mathbb{R}^{n_c \times p}$ and $\Pi \in \mathbb{R}^{n_c \times n_c}$ satisfying (1.60) are designed. The rows of the matrix U are denoted u_i.

2 A pair of feedback gains $K_y \in \mathbb{R}^{m \times p}$ and $K_z \in \mathbb{R}^{m \times n_c}$ relative to the system (4.80) is designed. It assigns several eigenvalues, the *corresponding eigenvalues and right eigenvectors are denoted λ_i and v_i.*

Then if U, T, Π, K_y and K_z are used as depicted in Figure 1.10 (page 55) the corresponding observer-based closed-loop system is such that

1. *The eigenvalues of Π belong to the observer-based closed-loop spectrum and the corresponding left eigenvectors are of the form $[u_i \; u_{ci}]$ for some vector $u_{ci}^* \in \mathbb{C}^{n_c}$.*

2. *The λ_i's belong to the observer-based closed-loop spectrum and the corresponding right eigenvectors are of the form $[v_i^* \; v_{ci}^*]^*$ for some vector $v_{ci} \in \mathbb{C}^{n_c}$.*

Proof. This result is already proved in part in Lemma 4.7.6, but for identifying the assigned eigenvectors, a more complete proof is proposed. The feedback $u = K_y y + K_z \hat{z}$ is applied to the system of Equation (4.79).

$$\begin{bmatrix} \dot{x} \\ \dot{\hat{z}} \end{bmatrix} = \begin{bmatrix} A & 0 \\ -TC & \Pi \end{bmatrix} + \begin{bmatrix} B \\ UB \end{bmatrix} \widetilde{K} \begin{bmatrix} C & 0 \\ 0 & I \end{bmatrix} \begin{bmatrix} x \\ \hat{z} \end{bmatrix}$$

where (see page 247)

$$\widetilde{K} = \begin{bmatrix} K_y & K_z \end{bmatrix} \left(I - \begin{bmatrix} D \\ 0 \end{bmatrix} \begin{bmatrix} K_y & K_z \end{bmatrix} \right)^{-1} \tag{4.81}$$

Multiplying on the left by

$$\begin{bmatrix} I & 0 \\ -U & I \end{bmatrix}$$

and on the right by

$$\begin{bmatrix} I & 0 \\ U & I \end{bmatrix}$$

using the identity $UA + TC = \Pi U$ (4.75) and denoting $\varepsilon = \hat{z} - Ux$

$$\begin{bmatrix} \dot{x} \\ \dot{\varepsilon} \end{bmatrix} = \begin{bmatrix} A & 0 \\ 0 & \Pi \end{bmatrix} + \begin{bmatrix} B \\ 0 \end{bmatrix} \widetilde{K} \begin{bmatrix} C & 0 \\ U & I \end{bmatrix} \begin{bmatrix} x \\ \varepsilon \end{bmatrix}$$

or

$$\begin{bmatrix} \dot{x} \\ \dot{\varepsilon} \end{bmatrix} = \begin{bmatrix} A + B\widetilde{K} \begin{bmatrix} C \\ U \end{bmatrix} & B\widetilde{K} \begin{bmatrix} 0 \\ I \end{bmatrix} \\ 0 & \Pi \end{bmatrix} \begin{bmatrix} x \\ \varepsilon \end{bmatrix} \tag{4.82}$$

The bloc triangular structure of this matrix leads to the property concerning eigenvalues that is stated in the lemma. Now, let us consider the right eigenvectors. If (λ_i, v_i) is assigned by K_y and K_z relative to System (4.80), it is also a pair of eigenvalues and eigenvectors relative to the matrix at the top left corner in (4.82). Therefore $[v_i^* \ 0]^*$ is an eigenvector relative to the system of Equation (4.82). Coming back to the natural basis, the assigned eigenvector is of the form

$$\begin{bmatrix} v_i \\ v_{ci} \end{bmatrix} = \begin{bmatrix} I & 0 \\ U & I \end{bmatrix} \begin{bmatrix} v_i \\ 0 \end{bmatrix} = \begin{bmatrix} v_i \\ Uv_i \end{bmatrix}$$

as stated in the lemma. A similar argument can be applied to left eigenvectors. ∎

Comment 1. The above result is not precisely the "Separation Principle" because the feedback design step depends on the knowledge of U. However, if the size of U is large enough so that the state vector can be recovered from y and z, the feedback design step can be considered as a state feedback design step that is really independent of the observer design step. In that case the "separation" is effective.

Comment 2. As discussed on page 64, we can ignore the lower part of assigned right eigenvectors (v_{ci}). For the same reason, the right hand side part of assigned left eigenvectors (u_{ci}) can also be ignored. So, observer based feedback design can be viewed as a left eigenvector assignment technique.

4.7.5 Equivalent dynamic controller

A technique for finding observer-based equivalent controllers can be found in [Alazard and Apkarian, 1999]. This technique considers observers with Kalman filter structure (see Figure 1.8 (page 51)), for this reason, the use of the Youla parameters becomes necessary when the controller order is not equal to the number of states of the system. The approach stated below considers banks of elementary observers (*i.e.*, general Luenberger observers without necessarily state vector estimation) giving more freedom as the respective sizes of the system and of the controller can be ignored.

Design Procedure *for finding an observer-based controller equivalent to a given one denoted* (A_c, B_c, C_c, D_c). The controller order will be denoted n_c $(A_c \in \mathbb{R}^{n_c \times n_c})$.

Step 1. Closed-loop system state-space matrix $\tilde{A}$. Compute $\Phi_1 = (I - D_c D)^{-1}$, $\Phi_2 = (I - D D_c)^{-1}$ and

$$\tilde{A} = \begin{bmatrix} A + B\Phi_1 D_c C & B\Phi_1 C_c \\ B_c \Phi_2 C & A_c + B_c \Phi_2 D C_c \end{bmatrix}$$

Step 2. Observer poles selection. Select n_c self-conjugate closed-loop eigenvalues of $\widetilde{A}$ that will be considered as observer poles and compute the corresponding left eigenvectors. These vectors are naturally partitioned under the form $[u_i \ u_{ci}]$ (u_i is a n-dimensional row vector, u_{ci} is a n_c-dimensional row vector). Define U, U_c and Π:

$$U = \begin{bmatrix} u_1 \\ \vdots \\ u_{n_c} \end{bmatrix} \ ; \ U_c = \begin{bmatrix} u_{c1} \\ \vdots \\ u_{cn_c} \end{bmatrix} \ ; \ \Pi = \mathrm{Diag}\{\pi_1, \dots \pi_{n_c}\}$$

Note that the matrix U_c must be invertible.

Step 3. Observer definition. Compute T as follows

$$T = UB\Phi_1 D_c + U_c B_c \Phi_2$$

the observer is defined by (U, T, Π), and by the feedback matrices K_y and K_z:

$$K_y = D_c$$
$$K_z = -C_c U_c^{-1}$$

Step 4. Feedforward definition. The feedforward state-space representation is

$$A_f' = \Pi \ ; \ B_f' = UB + TD \ ; \ C_f' = C_c U_c^{-1} \ ; \ D_f' = I$$

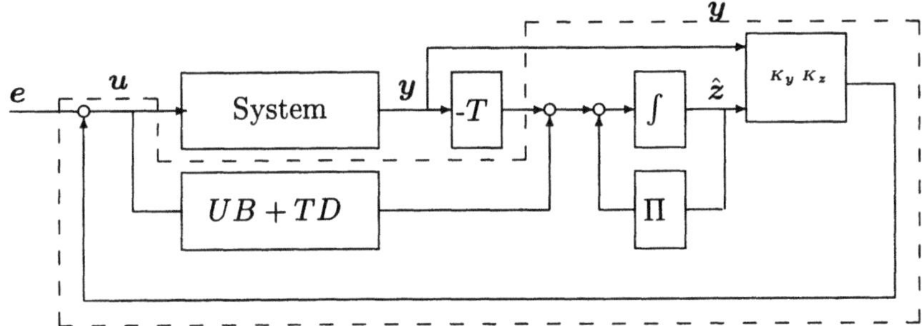

Feedforward to be cancelled

Figure 4.2. Observer-based controller.

Justification. The proposed proof consists of showing that triangularizing the closed-loop matrix $\widetilde{A}$, we obtain the block triangular form of the Separation Principle subsection. The matrix $\widetilde{A}$ of Step 1 is the state-space matrix obtained by closure of the loop relative to the system

(A, B, C, D) and to the controller (A_c, B_c, C_c, D_c). First, this matrix is block-diagonalized. The change of basis used for that purpose is

$$\begin{bmatrix} I & 0 \\ -U & -U_c \end{bmatrix} \quad \text{with inverse} \quad \begin{bmatrix} I & 0 \\ -U_c^{-1}U & -U_c^{-1} \end{bmatrix}$$

$\widetilde{A}$ becomes

$$\begin{bmatrix} A + B\Phi_1 D_c C - B\Phi_1 C_c U_c^{-1} U & -B\Phi_1 C_c U_c^{-1} \\ 0 & \Pi \end{bmatrix} \tag{4.83}$$

It remains to identify the sub-blocks in Equations (4.82) and (4.83). For that purpose, $\widetilde{K}$ (see Equation (4.81)) can be written

$$\widetilde{K} = \begin{bmatrix} (I - K_y D)^{-1} K_y & (I - K_y D)^{-1} K_z \end{bmatrix}$$

therefore, the state-space matrix of Equation (4.82) can be written

$$\begin{bmatrix} A + B(I - K_y D)^{-1} K_y C + B(I - K_y D)^{-1} K_z U & B(I - K_y D)^{-1} K_z \\ 0 & \Pi \end{bmatrix}$$

that is, after substitution of $K_y = D_c$ and $K_z = -C_c U_c^{-1}$ and $(I - K_y D)^{-1} = \phi_1$, the block triangular form of $\widetilde{A}$ of Equation (4.83). It remains to check that the matrices T, U and Π correspond to an observer (see (4.75)). For that purpose, it suffices to note that the zero block in (4.83) corresponds to $U(A + B\phi_1 D_c C) + U_c B_c \phi_2 C = \Pi U$ that can be written $UA + TC = \Pi U$ using the form of T given in the design procedure.

Concerning the feedforward gain, from Figure 4.2, the observer structure introduces a feedforward gain (given by Equation (1.76), page 56) that must be cancelled. It is straightforward to check that (A'_f, B'_f, C'_f, D'_f) is the dynamic inverse of (A_f, B_f, C_f, D_f) of Equation (1.76). ∎

Notes

1 It is not necessary to fix the denominator coefficients for obtaining constraints that are linear with respect to the coefficients b_{ijk} *and* a_{ijk}. For example if all denominators are identical, multiply (4.12) by the common denominator. But the constraints defined here will be combined to other ones that would not be linear without fixing the coefficients a_{ijk}.

2 If $\xi = 0$ and $x \neq 0$, it means that β is a non observable eigenvalue, which is anyway "assigned". This is the observability criterion mentioned on page 247.

3 In the output feedback case, denoting m_i the number of poles to be associated with the ith regulated output, we must have $n_z + \sum m_i = p$ ($n - p$ ignored non-dominant poles).

Chapter 5

APPENDIX 2: ADDITIONAL TOPICS

5.1 Models used for demonstrations

The following models are considered for demonstration

- A simplified lateral model of a flexible aircraft.

- A bank of longitudinal and lateral linearized models of the RCAM (Research Civil Aircraft Model, [Magni et al., 1997b]).

5.1.1 A simplified flexible aircraft

5.1.1.1 System 1

System 1 is a linear model of the lateral motion of an aircraft with states $\{\beta, p, r, \phi, x_5, x_6\}$. These states are respectively the angle of sideslip, roll rate, yaw rate, roll angle, x_5 and x_6 correspond to actuators in series on both inputs. The first input is the aileron deflection (deg), the second is the rudder defection (deg). The measurements are $\{n_y$ (lateral load factor) $, p, r, \phi\}$ with respective units g, deg/s, deg/s, deg.

This system is obtained as follows: `sys = demodata(1);`

As this system is used for "academic" illustration, design specifications are not given.

5.1.1.2 System 2

System 2 is similar to System 1 but with six additional states modeling the flexibility of the aircraft.

This system is obtained as follows: `sys = demodata(2);`

5.1.2   A bank of linearized models of an aircraft (the RCAM)

We shall consider a set of 37 linearized models of a large transportation aircraft: the RCAM. These models are given for the longitudinal and the lateral motion. They were obtained by linearizing the nonlinear model described in [Lambrechts et al., 1997]. This set of models covers all the flight envelope to be considered during the landing phase. These models are generated by calling `rcamdata`.

5.1.2.1   Longitudinal motion

States, outputs, inputs (international system units):

- States vector: the first four states are the pitch rate q, pitch angle θ, forward velocity in body-axis u_B, vertical velocity in body-axis w_B.

- Measurement vector: vertical velocity in vehicle axis w_V, total velocity V, pitch rate q, vertical load factor n_z.

- Input vector: tailplane and engine commands.

Simplified set of *performance* objectives:

- Settling time for $V < 45$s.

- Settling time for $w_V < 20$s.

- Peak value of V less than 1 for a demand of w_V equal to 4.2.

- Peak value of w_V less than 0.7 for a demand of V equal to 13.

- Overshoot less than 1%.

Robustness: the above performance requirement should be met for

- mass variations from 100000 to 150000.

- total velocity variations (depends on the mass).

- horizontal position of the center of gravity (from 0.15% to 0.31%).

- vertical position of the center of gravity (from 0.00% to 0.21%).

- delay form 0.05 to 0.1.

The set of models obtained by calling `rcamdata` describe all this flight envelope.

Other requirements: to insure comfort and low control activity, the controller must be as "slow" as possible within the limits fixed by the settling time requirements. High pass effect of gains should be avoided as much as possible.

5.1.2.2   Lateral motion

States, outputs, inputs (international system units):

- States vector: the first four states are the roll rate p, yaw rate r, roll angle ϕ, lateral velocity in the body-axis....

- Measurement vector: angle of side slip β, roll rate p, yaw rate r, roll angle ϕ.

- Input vector: aileron and rudder commands.

Simplified set of *performance* objectives:

- Settling time for $\phi < 8$s.

- Settling time for $\beta < 10$s.

- Peak value of ϕ less than 1 for a demand of β equal to 2.

- Peak value of β less than 1 for a demand of ϕ equal to 20.

- Overshoot less than 1%.

Robustness: similar to the longitudinal case.

Other requirements: to insure comfort and low control surface activity, the controller must be as "slow" as possible within the limits fixed by the settling time requirements. High pass effect of gains should be avoided as much as possible.

5.1.2.3   Software

In the lateral case, these linearized models "**sys**" are obtained as follows:

```
[sys,b2,d2] = rcamdata('lat',model_number,option,delay)
```

In the longitudinal case:

```
[sys,b2,d2] = rcamdata('lon',model_number,option,delay)
```

Information on the model numbers can be found in Table 5.1. The output arguments b2 and d2 will be defined later (see, architecture with integrators). The input argument **option** permits us to add actuator models (**option** = 1) and integrators (**option** = 2). Two specific analysis tools are also provided:

- **rcampole** to plot the pole map of all linearized models

- **rcamstep** to plot the step responses of all linearized models.

5.1.2.4   Feedback architecture

For both lateral and longitudinal cases, two control architectures are of interest. Let us define the regulated output z and the corresponding reference signals z_R. In the lateral case:

$$z = \begin{bmatrix} \phi \\ \beta \end{bmatrix} \;\; ; \;\; z_R = \begin{bmatrix} \phi_R \\ \beta_R \end{bmatrix}$$

in the longitudinal case

$$z = \begin{bmatrix} V \\ v_w \end{bmatrix} \;\; ; \;\; z_R = \begin{bmatrix} V_R \\ v_{wR} \end{bmatrix}$$

Architecture without integrators. It corresponds to the models obtained with `option = 0 or 1`. The feedback we look for is

$$u = Ky + Hz_R$$

Both (possibly dynamic) gains H and K are to be designed.
Architecture with integrators. It corresponds to Figure 5.1. The feedback we look for (with possibly dynamic gains) is

$$u = K_1 y + K_2 \int (z - z_R)$$

The measures are y and $\int (z - z_R)$ and the integrators states are added

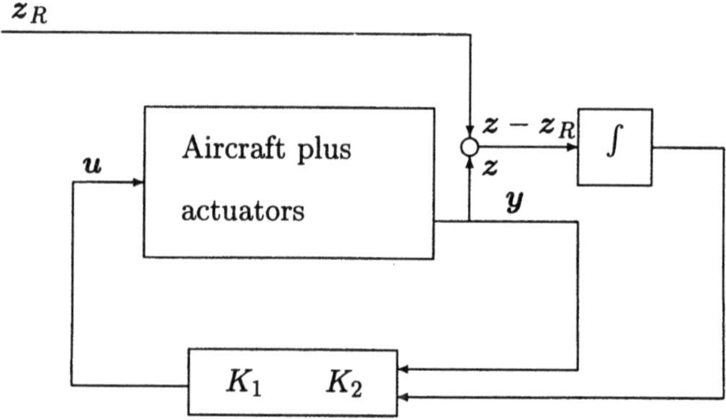

Figure 5.1.   Architecture with integrators

to the natural aircraft states. This augmented system is the first output argument (**sys**) when **rcamdata** is invoked with **option** = 2. Two additional output arguments **b2** and **d2** are the "B" and "D-matrices" corresponding to the reference input z_R when the feedback loop is closed.

Model description							
Number	Mass	CoG hori	CoG vert	Air speed	Delay	Pole symbols	Step symbols
0	120	0.23	0	80	0.075	rd	r-
1	150	0.15	0	80	0.075	yo	y-
2	150	0.15	0.21	80	0.075	y+	y:
3	150	0.31	0	80	0.075	yx	y-.
4	150	0.31	0.21	80	0.075	y^*	y- -
5	100	0.15	0	80	0.075	go	g-
6	100	0.15	0.21	80	0.075	g+	g:
7	100	0.31	0	80	0.075	gx	g-.
8	100	0.31	0.21	80	0.075	g^*	g- -
9	150	0.15	0	70	0.075	mo	m-
10	150	0.15	0.21	70	0.075	m+	m:
11	150	0.31	0	70	0.075	mx	m-.
12	150	0.31	0.21	70	0.075	m^*	m- -
13	150	0.15	0	90	0.075	bo	b-
14	150	0.15	0.21	90	0.075	b+	b:
15	150	0.31	0	90	0.075	bx	b-.
16	150	0.31	0.21	90	0.075	b^*	b- -
17	120	0.15	0	80	0.05	c+	c-
18	120	0.31	0	80	0.05	cx	c:
19	120	0.23	0	80	0.05	co	c-.
20	120	0.23	0	80	0.1	c^*	c- -
21	100	0.15	0	60	0.075	ko	w-
22	100	0.15	0.21	60	0.075	k+	w:
23	100	0.31	0	60	0.075	kx	w-.
24	100	0.31	0.21	60	0.075	ks	w- -
25	100	0.15	0	90	0.075	ro	r-
26	100	0.15	0.21	90	0.075	r+	r:
27	100	0.31	0	90	0.075	rx	r-.
28	100	0.31	0.21	90	0.075	r^*	r- -
29	120	0.23	0	60	0.075	kd	y- -
30	120	0.23	0	65	0.075	bx	g-.
31	120	0.23	0	70	0.075	y^*	m:
32	120	0.23	0	75	0.075	co	b-
33	120	0.23	0	85	0.075	m+	c- -
34	120	0.23	0	90	0.075	gx	w-.
35	150	0.31	0.21	75	0.075	kd	y- -
36	150	0.31	0.21	85	0.075	c^*	g-.

Table 5.1.   Bank of linearized models of the RCAM

5.2 Matrices CSTR and CRIT

This section gives information which is not necessary for a standard use of the toolbox. The functions fb_dyn and fb_tun handle two matrices CRIT and CSTR. These matrices contain the following information:

Structure of CSTR. CSTR is a matrix defined as follows

$$
\mathrm{CSTR} = \begin{bmatrix}
[\text{ N1 } \text{ N2 } \text{ opt } \text{ m } \text{ p } \text{ n1 } \text{ n2 } \text{ N10 } \text{ N20}\dots\ ] \\
[\text{ A } \text{ B }] \\
\text{Row}(\text{Matdeg})\ 0\dots0 \\
\text{Row}(\text{Dmin})\ 0\dots0 \\
\text{Row}(\text{Dmax})\ 0\dots0 \\
\text{Row}(\text{Nmin})\ 0\dots0 \\
\text{Row}(\text{Nmax})\ 0\dots0
\end{bmatrix}
$$

- The matrices A and B define the linear constraints of the Linear Quadratic Problems stated in Lemmas 4.1.4 and 4.2.4. The N1 first rows of A and B correspond to the equality constraints (A_1 and b_1), the N2 additional rows correspond to the inequality constraints (A_2 and b_2).

- m, p are the numbers of inputs and outputs of the considered system (converse for the feedback gain).

- n1 is the number of columns of [A   B] and n2 is the number of columns of Row(Dmin) ("Row(A)" stands for the row vector consisting of all the rows of the matrix A in natural order).

- N1, N2, N10, N20 permit the designer to identify the last set of constraints that was added into CSTR: The last time this matrix was modified, the rows N10+1 to N1 and N20+1 to N2 were added to [A B].

- The matrices Matdeg, Dmin, Dmax, Nmin, Nmax define the constraints on the gain structure with good "visibility". The precise use of these matrices can be found by typing help ktf_cstr. Using this function the user might define customized constraints relative to the gain structure (replacing the use of str_cstr).

Structure of CRIT. This matrix contains a quadratic criterion $J = 0.5x^T \mathrm{H} x + \mathrm{F} x$

$$
\mathrm{CRIT} = \begin{bmatrix} \mathrm{H} \\ \mathrm{F} \end{bmatrix}
$$

In fact, a constant component of the criterion is not considered, the actual value of the criterion is

$$
J = 0.5x^T \mathrm{H} x + \mathrm{F} x + 0.5x \mathrm{H}^{-1} x
$$

Note that (using natural notation) CRIT = Q1 * CRIT1 + Q2 * CRIT2 corresponds to $J = Q_1 J_1 + Q_2 J_2$.

5.3 Installation and system requirements

Installation. The functions provided in this toolbox should be placed in a directory `rmct` placed in the same location as the rest of MATLAB files. Set up the MATLABPATH environment variable to include this directory. See also the `README.txt` file.

System requirements. It is assumed that Version 5 or higher of MATLAB is available. The Optimization Toolbox and the Control System Toolbox are also required.

The Control System Toolbox is required only for two functions (`lsim_mod` and `fb_dyn`). But the Control System Toolbox is used in most illustrative examples.

The function `quadprog` of the Optimization Toolbox (version 2.0 or posterior) is used in two of the main functions that are `fb_dyn` and `fb_tun`. The function `fminunc` of the Optimization Toolbox is used in `azer`, `kern_ins`, `ob_ins`, `sfb_ins`, `dfb_ins` and `smalldfb`. (See the `README.txt` file for use with a version of the Optimization Toolbox anterior to 2.0).

Conclusion

A selection of modal control design techniques were presented including some original contributions ,as eigenvector selection, observer design and dynamic feedforward design.

Our main contribution is the multi-model control design technique that is presented in §1.3. Robustness against *real* uncertainties can be treated with this technique in a very efficient way *i.e.*, without conservatism and with controllers of reasonable order. This technique, originally developed in a *multi-model* setting, can also be used for designing dynamic controllers relative to a *single* model, because in this case it is also useful to consider structured controllers, frequency domain constraints and phase control. Single-model controller order reduction is an additional advantage of this approach. Note that the multi-model control design technique can easily be initialized by a given controller, so, if this controller was designed using a frequency domain approach, it becomes possible to combine the complementary advantages of modal and frequency domain approaches.

The illustrative examples proposed in this book are quite simple, for a more realistic illustration of the potentialities of modal control the reader is referred to [Chiappa et al., 1998, Chiappa et al., 2001b, Döll et al., 2000, Le Gorrec et al., 1998a, Le Gorrec et al., 1998b]

From the theoretical point of view, the contents of this book can already be extended as follows:

- by considering LMIs (Linear Matrix Inequalities) instead of the LQP (Linear Quadratic Programming) approach, see [Magni et al., 1998, Le Gorrec., 1998] for details.

- by using simultaneously μ-analysis and multi-model control design. μ-analysis is used for identifying worst cases (in this book, worst case analysis was limited to standard analysis relative to a bank of models). The advantage of using μ-analysis, is that worst cases cannot be missed as not belonging to the considered bank. See [Magni et al., 1998] for details.

- by designing directly scheduled controllers, see [Magni, 1999, Döll et al., 2000] for details. The approach proposed in these references consists of choosing *a priori* a symbolic form of the scheduled gains, then the models are accordingly modified in such a way that multi-model control design can be applied ignoring scheduling.

- by considering high gain feedback theory ([Willems, 1981]). It is possible to interpret this theory in terms of asymptotic eigenstructure assignment, however, practical applications are not as common as with standard eigenstructure assignment (examples of applications: asymptotic Loop Transfer Recovery, singular $\mathcal{H}_\infty$ problems).

Using models in LFT (Linear Fractional Transformation) form is a key point for future developments. A toolbox for low order LFT modeling is being developed, so that new techniques using both multi-model modal control and LFT manipulation can be considered:

- Most equations encountered in §1.2 can be written in terms of LFTs, and solutions can also be found in terms of LFTs. This remark justifies a new approach to modal control in which models, controllers and performance are expressed as LFTs. Note that controllers in LFT form correspond to gain scheduling. It is also quite natural to consider modal performance (*e.g.*, pole and eigenvector assignment) in LFT form. For example, assume that some assigned poles are more or less fast depending on a mass or an inertia, in this case the assigned poles can be viewed as scalar LFTs.

- In the same spirit, a given nominal controller can be automatically transformed into an LFT form, preserving in that way the nominal features when some parameters (those that can be measured) vary. For that purpose the first step consists of obtaining an equivalent modal version of the given controller (*e.g.*, using the design procedure of §4.7.5 or the technique of §2.6.2). The second step consists of "preserving this assignment" by replacing matrices by LFTs in the equivalent modal version.

All systems are specific, so we do not believe in modal control routines that work for all kinds of systems (keep in mind the warnings given in §2.1.1 concerning the standalone functions of the toolbox). Nevertheless, we plan to develop some additional functions expected to work for large classes of systems. Two projects are considered:

- systematic controller order reduction.

- automatic handling of the iterations alternating μ-analysis and multi-model control design for systems in LFT form. This technique is reminiscent of (D, K)-iterations but should be efficient in the case of real uncertainties.

References

[Ackermann, 1993] Ackermann, J. (1993). *Robust Control system with uncertain physical parameters*. Communication and Control Engineering Series. Springer-Verlag, London.

[Alazard and Apkarian, 1999] Alazard, D. and Apkarian, P. (1999). Exact observer based structure for arbitrary compensator. *Int. J. Robust and Nonlinear Control*, 9(2):101–118.

[Apkarian et al., 1989] Apkarian, P., Champetier, C., and Magni, J. (1989). Design of a helicopter output feedback control law using modal and structured-robustness techniques. *Int. J. Control*, 50(4):1195–1215.

[Barmish, 1994] Barmish, B. (1994). *New tools for robustness of linear systems*. MacMillan Publishing Company.

[Champetier and Magni, 1991] Champetier, C. and Magni, J. (1991). On eigenstructure assignment by gain output feedback. *SIAM J. Contr. Optimiz.*, 29(4):848–865.

[Chiappa et al., 1998] Chiappa, C., Magni, J., , Döll, C., and Le Gorrec, Y. (1998). Improvement of the robustness of an aircraft autopilot designed by an H_∞ technique. *In Proc. CESA'98 Conference, Nabeul-Hammamet, Tunisia*, 1:1016–1020.

[Chiappa et al., 2001a] Chiappa, C., Magni, J., and Le Gorrec, Y. (2001a). Méthodologies de synthèse de commandes de vol d'un avion souple. *Journal Européen des systèmes automatisés*, 35(1-2):191–208.

[Chiappa et al., 2001b] Chiappa, C., Magni, J., and Le Gorrec, Y. (2001b). A modal multimodel approach for controller order reduction and structuration. *IEEE 6th Conference on Control Applications, Mexico*.

[De Larminat and Houisot, 1993] De Larminat, P. and Houisot, P. (1993). Application of ACSYDE to the IFAC-93 benchmark. *In Proc. IFAC world Congress, Sydney, Australia*, 3:177–180.

[Döll et al., 2000] Döll, C., Le Gorrec, Y., Magni, J., and Ferreres, G. (2000). Design of a robust self-scheduled missile autopilot by multi-model eigenstructure assignment. *In Proc. Asian Control Conference, Shangaï, China*.

[Doyle et al., 1989] Doyle, J., Glover, K., Khargonekar, P., and B.C., F. (1989). State-space solutions to standard H_2 and H_∞ control problems. *IEEE Transactions on Automatic Control*, AC-34(8):831–846.

[Doyle and Stein, 1979] Doyle, J. and Stein, G. (1979). Robustness with observers. *IEEE Transactions on Automatic Control*, AC-24:607–611.

[Doyle and Stein, 1981] Doyle, J. and Stein, G. (1981). Multivariable feedback designs : Concepts for a classical/modern synthesis. *IEEE Transactions on Automatic Control*, AC-26:4–16.

[Falb and Wolovich, 1967] Falb, P. and Wolovich, W. (1967). Decoulping in the design and synthesis of multivariable control systems. *IEEE Transactions on Automatic Control*, AC-12:651–659.

[Farineau, 1989] Farineau, J. (1989). Lateral electric flight control laws of the a320 based upon eigenstructure assignment techniques. *In Proc. AIAA Conf. on Guidance Navigation and Control, Boston*, pages 1367–1372.

[Garcia et al., 1997] Garcia, G., Bernussou, J., Daafouz, J., and Arzelier, D. (1997). Robust quadratic stabilization, Tutorial Part. *In "Robust Flight Control" Lecture Note in Control and Information Sciences Springer-Verlag*, 224:42–51.

[Gilbert, 1969] Gilbert, E. G. (1969). The decoupling of multivariable system by state feedback. *SIAM J. Contr. Optimiz.*, 7(1):50–63.

[Hamdan and Nayfeh, 1989] Hamdan, A. and Nayfeh, A. (1989). Mesures of modal controllability and observability for first- and second-order linear systems. *Journal of Guid. Contr. and Dyn.*, 12(3):421–428.

[Horowitz, 1991] Horowitz, I. (1991). Survey of quantitative feedback theory. *Int. J. Control*, 53(2):255–291.

[Joos, 1997] Joos, H. (1997). Multi-objective parameter synthesis (MOPS), RCAM Part. *In "Robust Flight Control" Lecture Note in Control and Information Sciences Springer-Verlag*, 224:199–217.

[Joos, 1999] Joos, H. (1999). A methodology for multi-objective design assessment and flight control synthesis tuning. *Aerospace Science and Technology*, 3(3):161–176.

[Junkins and Kim, 1993] Junkins, J. and Kim, Y. (1993). *Introduction to dynamics and control of flexible structures*. AIAA Education Series. AIAA.

[Kaustky and Nichols, 1990] Kaustky, J. and Nichols, N. (1990). Robust pole assignment in systems subject to structured perturbations. *Systems and Control Letters*, 15:373–380.

[Kaustky et al., 1985] Kaustky, J., Nichols, N., and Van Dooren, P. (1985). Robust pole assignment in linear state feedback. *Int. J. Control*, 41(5):1129–1155.

[Khargonekar et al., 1990] Khargonekar, P., Petersen, I., and Zhou, K. (1990). Robust stabilization of uncertain systems: Quadratic stability and H∞ control theory. *IEEE Transactions on Automatic Control*, 35:356–361.

[Kimura, 1975] Kimura, H. (1975). Pole assignment by gain output feedback. *IEEE Transactions on Automatic Control*, AC-20:509–516.

[Lambrechts et al., 1997] Lambrechts, P., Bennani, S., Looye, G., and Moormann, D. (1997). Definition of the rcam design challenge problem. *In "Robust Flight Control" Lecture Note in Control and Information Sciences Springer-Verlag*, 224:149–179.

[Laub and Moore, 1978] Laub, A. and Moore, B. (1978). Calculation of transmission zeros uzing QZ techniques. *Automatica*, 14:557–566.

[Le Gorrec., 1998] Le Gorrec., Y. (1998). Commande modale robuste, synthèse de gains auto-séquencés: approche multimodèle. *Thèse présentée à l'Ecole Nationale Supérieure de l'Aéronautique et de l'Espace (SUPAERO), Toulouse, France*.

[Le Gorrec et al., 1997] Le Gorrec, Y., J.F., M., and Chiappa, C. (1997). Flexible transmission system controlled by modal dynamic feedback. *European Journal of Control*, 3:227–234.

[Le Gorrec et al., 1998a] Le Gorrec, Y., Magni, J., Chiappa, C., and Kubica, F. (1998a). Structured gain design applied to aircarft autopilot design. *In Proc. CESA'98 Conference, Nabeul-Hammamet, Tunisia*, 1:1011–1015.

[Le Gorrec et al., 1998b] Le Gorrec, Y., Magni, J., Döll, C., and Chiappa, C. (1998b). A modal multimodel control design approach applied to aircraft autopilot design. *AIAA Journal of Guidance, Control, and Dynamics*, 21(1):77–83.

[Liu and Patton, 1998] Liu, G. and Patton, R. (1998). *Eigenstructure Assignment for Control System Design*. Wiley, London.

[Livet, 1995] Livet, T. (1995). Commande robuste des avions par les techniques modales. *Thèse présentée à l'Ecole Nationale Supérieure de l'Aéronautique et de l'Espace (SUPAERO), Toulouse, France*.

[Luenberger, 1966] Luenberger, D. (1966). Observers for multivariable systems. *IEEE Transactions on Automatic Control*, AC-11(2):190–197.

[MacFarlane and Karcanias, 1976] MacFarlane, A. and Karcanias, K. (1976). Poles and zeros of linear multivariable systems : a survey of the algebraic, geometric and complex-variable theory. *Int. J. Control*, 24(1):33–74.

[Maciejowski, 1989] Maciejowski, J. (1989). *Multivariable Feedback Design*. Electronic Systems Engineering Series. Addisson-Wesley Publishing Company, Wokingham, England.

[Magni, 1987] Magni, J. (1987). Commande modale des systèmes multivariables. *D.Sc. Thesis No. 1316, Université Paul Sabatier, Toulouse, France*.

[Magni, 1990] Magni, J. (1990). Observer synthesis by eigenstructure assignment - an application to "Loop Transfer Recovery". *In Proc. of the 11th Triennial IFAC World Congress, Tallin*, II:155–168.

[Magni, 1996] Magni, J. (1996). Continuous time parameter identification by using observers. *IEEE Transactions on Automatic Control*, AC-40(10):1789–1792.

[Magni, 1999] Magni, J. (1999). Multimodel eigenstructure assignment in flight-control design. *Aerospace Science and Technology*, 3(3):141–151.

[Magni and Champetier, 1991] Magni, J. and Champetier, C. (1991). A geometric framework for pole assignment algorithms. *IEEE Trans. Autom. Control*, AC-36:1105–1111.

[Magni et al., 1991] Magni, J., Champetier, C., and Ménard, P. (1991). A new tool for the analysis of modal control laws : The pole attractors. *IEEE Transactions on Automatic Control*, AC-36(2):219–223.

[Magni et al., 1998] Magni, J., Le Gorrec, Y., and Chiappa, C. (1998). A multimodel-based approach to robust and self-scheduled control design. *In Proc. 37th I.E.E.E. Conf. Decision Contr., Tampa, Florida*, pages 3009–3014.

[Magni et al., 1997a] Magni, J., Le Gorrec, Y., Chiappa, C., and Alazard, D. (1997a). Flexible structure control by eigenstructure assignment. *In Proc. of the IFAC-IFIP-IMACS Conference on Control and Industrial Systems, Belfort, France*, 2:99–104.

[Magni and Manouan, 1994] Magni, J. and Manouan, A. (1994). Robust flight control design by eigenstructure assignment. *In Proc. of the IFAC Symposium on Robust Control, Rio de Janeiro, Brasil*, pages 388–393.

[Magni and Mouyon, 1991] Magni, J. and Mouyon, P. (1991). A tutorial approach to observer design. *In Proc. AIAA Conf. on Guidance Navigation and Control, New Orleans*, III:1748–1755.

[Magni and Mouyon, 1994] Magni, J. and Mouyon, P. (1994). On residual generation by observer and parity space approaches. *IEEE Transactions on Automatic Control*, 39(2):441–447.

[Magni et al., 1997b] Magni, J., Terlouw, J., and Bennani (Eds.), S. (1997b). *Robust Flight Control*. Lecture Notes in Control and Information Sciences, No 224. Springer-Verlag.

[McFarlane and Glover, 1990] McFarlane, D. and Glover, K. (1990). *Robust Controller Design Using Normalized Coprime Factor Plant Descriptions*. Lecture Notes in Control and Information Sciences. Springer-Verlag.

[Moore, 1976] Moore, B. (1976). On the flexibility offered by state feedback in multivariable system beyond closed loop eigenvalue assignment. *IEEE Transactions on Automatic Control*, AC-21:659–692.

[Moore, 1981] Moore, B. (1981). Principal component analysis in linear systems: Controllability, observability, and model reduction. *IEEE Transactions on Automatic Control*, AC-26:17–32.

[Moore and Klein, 1976] Moore, B. and Klein, G. (1976). Eigenvalue selection in the linear regulator combining modal and optimal control. *In Proc. 15th I.E.E.E. Conf. Decision Contr.*, pages 214–215.

[Rosenbrok, 1974] Rosenbrok, H. (1974). *Computed-Aided Control System Design*. Academic Press.

[Wilkinson, 1965] Wilkinson, J. H. (1965). *The Algebraic Eigenvalue Problem*. Monographs on Numerical Analysis. Clarendon Press, Oxford.

[Willems, 1981] Willems, J. (1981). Almost invariant subspaces : an approach to high gain feedback design. part1 : almost controlled invariant subspaces. *IEEE Transactions on Automatic Control*, AC-26:235–252.

[Wonham, 1970] Wonham, W. (1970). Dynamic observers-geometric theory. *IEEE Transactions on Automatic Control*, AC-15:258–259.

[Wonham, 1979] Wonham, W. (1979). *Linear Multivariable Control, a Geometric Approach*. Springer-Verlag, New York Heidelberg Berlin.

[Young, 1996] Young, P. (1996). Controller design with parametric uncertainty. *Int. J. Control*, 65(3):469–509.

[Zhou and Doyle, 1999] Zhou, K. and Doyle, J. (1999). *Essentials of Robust Control*. Prentice Hall Inc.

About the Author

The author is "Directeur de Recherche" at ONERA (Office National d'Etudes et de Recherches Aerospatiales) and Professor of Automatic Control at SUPAERO (French National College for Aerospace). He received his M.S. degree from Imperial College of the University of London, England and his "Docteur es Sciences" degree from Toulouse University, France. He has co-edited the book "Robust Flight Control: A Design Challenge" (Springer Verlag, LNCIS 224) and authored the book/toolbox "Linear Fractional Representation Toolbox" available on the web. His former research interests were in the areas of pole assignment by output feedback, theories related to the Geometric Approach, high gain feedback and fault diagnosis. His present research interests lie in the area of robust multimodel and self-scheduled control, robustness analysis and low order modelling in LFT form. The author was recipient of the "Best Application Paper Prize" at the IFAC World Congress, Beijing 1999.

Index

almost zeros, 102, **178**, **228**
assignable eigenvectors, 15, 65
bank of models, 294, **229**
cancellation (pole/zero), 43, 134, 214, 274
coherency of constraints, 111
combination of criteria, 82, 296
complementarity with $\mathcal{H}_\infty$, μ-synthesis, 72
complex values, 30, 49
constraints on
    eigenvector assignment, *see* options
    eigenvector tuning, 91, 256
    eigenvalue assignment, 33, 118-125, 190
    gain coefficients, 86, 67, 160, **192**, 240
    pole motion, 83, 70, 110, 153, **186**, 204
constraints
    global, 67, 68, 69, 86
    local, 70, 82
contollability
    analysis, 157, **224**
    for cancellation, 274
    definition, 247
    degree of, 100, 109, 157, **224**
    observer poles, 48
criterion for
    insensitivity, 35, 122
    overspecified problems, **183**
    underspecified problems, 29, 66, 81, 185, 215, 295
    minimum energy, 24
criterion (numerical limitation), **215**, 254
dominant mode, 95, 110, **217**, **226**
decomposition (modal), 64, 96, **217**
decoupling
    principle, 19, 37
    examples, 130-137

degrees of freedom, 16, 252, 267
direct transmission, 14, 49, 59, 247
dynamic feedback gains, 56, 45, 65, **175**, 208, 223, 249, 284
eigenstructure
    assignment, **195**, 188, 206, 209, 219
    decoupling, 19
    definition, 14
    minimum energy, 24, 262
    projection, 23
    pseudo-assignment, 276
eigenvalue
    assignment, 32, 190, **209**
    settling time, 119
    selection, 103, **239**, **243**
    sensitivity, 35, 71, **255**, 259, 198, 221, 233
    trace constraint, 111
feedback gain
    computation, 28, 57, 65 **208**, **211**, 209
    reduction, 72, 159
    structure, 67, 86, 160, 240
    templates, 68, 76, **174**
feedforward gain
    constant, 37, 40, **214**
    dynamic, 42, 134, **213**, 274
flexible system control, 70, 149
first order perturbations, 34, 83, 181, 255, 257, 259
hypothesis, 15, 245
frequency domain constraints, 68, **174**
input directions, 14
installation (toolbox), 297
Kalman structure, 51, 120, 123, 280,
loop transfer recovery
    definition, 60
    example, 137
linear quadratic progamming,

eigenstructure assignment, 67, 251
phase control, 70, 260
tuning, 81,
toolbox, 297
linear quadratic regulator, 262
LMI approach, 175
matrix
notation, 15, 48
identities, 247
m-files (examples of) for
multimodel, 73, 74, 142, 152, 153, 160, 162
observer design, 58, 137
tuning, 88, 145, 155, 187, 189, 191, 194
minimum energy, 24
modes
definition, 17
analysis, 96, 98, 100, **217**, **226**
multi-model
coherency, 111
CRIT, 67, 295
CSTR, 67, 295
dynamic gain, 70, 73, 74
proportional gain, 140
by tuning, 145
noninteractive control
feedforward-based, 134, **213**
observer-based, 135
standard, 37, 129, 268
notations, 12
observers
connection to systems, 55, **176**
elementary, 47, **219**, **278**
definition of U,T,Π, 48
definition of Q_y,Q_z, 49
design, 53, **219**, **221**, **237**
general, 48
minimum order, 50
less than minimum order, 57
state observers, 50, 58, **237**
observer-based feedback, 54, 57, 120 135, 137, **199**, **198**, **201**, **284**

observer-based state feedback, 58, 58
options for
decoupling, 22
desired vectors, 23
minimun energy, 25
pole assignment, 33
projections, 23
order reduction, 72, 162
output directions, 14, 47
phase control, 70, 74 153, **204**
pole assignment, 32, 265
recommendations
decoupling, 42, 273
general, 30, 53, 109
multi-model, 111, 114
tuning, 112
nominal design, 113
design cycle, 113, 115
repeated eigenvalues, 15, 111
residuals, 98, **226**
RCAM, **229**, **231**, **232**, **290**, **294**
sensitivity of (first order)
eigenvalues, 34, 71, 83, 122
eigenstructure, 90, **181**
separation principle, 57, **282**
settling time, 119
simulation, 96, **217**
state feedback, 28, 34,
tuning
eigenvectors, 90, **181**
dynamic controllers, 124, **175**
CRIT, 82, 295
CSTR, 82, 295
initialization of, 86, 113
observer-based contr., 126, **176**
principle, 80, 88, **211**
under specified assignment, see criteria
unknown input observer, 52, 60, 137, **219**
zeros,
computation, 102, **178**
decoupling, 42, 61, 266
definition, 270
zero directions, 103, **195**